AF796383

Topological States for New Modes of Information Storage and Transfer

Prabhakar Bandaru • Shreyam Natani

Topological States for New Modes of Information Storage and Transfer

A Survey of the Status of the Discovery of the Majorana Modes in the Solid State

Springer

Prabhakar Bandaru
Department of Mechanical & Aerospace Engineering, Department of Electrical Engineering, Materials Science Program
University of California, San Diego
La Jolla, CA, USA

Shreyam Natani
Mechanical & Aerospace Engineering
University of California, San Diego
San Diego, CA, USA

ISBN 978-3-030-93339-5 ISBN 978-3-030-93340-1 (eBook)
https://doi.org/10.1007/978-3-030-93340-1

© The Editor(s) (if applicable) and The Author(s), under exclusive license to Springer Nature Switzerland AG 2022
This work is subject to copyright. All rights are solely and exclusively licensed by the Publisher, whether the whole or part of the material is concerned, specifically the rights of translation, reprinting, reuse of illustrations, recitation, broadcasting, reproduction on microfilms or in any other physical way, and transmission or information storage and retrieval, electronic adaptation, computer software, or by similar or dissimilar methodology now known or hereafter developed.
The use of general descriptive names, registered names, trademarks, service marks, etc. in this publication does not imply, even in the absence of a specific statement, that such names are exempt from the relevant protective laws and regulations and therefore free for general use.
The publisher, the authors and the editors are safe to assume that the advice and information in this book are believed to be true and accurate at the date of publication. Neither the publisher nor the authors or the editors give a warranty, expressed or implied, with respect to the material contained herein or for any errors or omissions that may have been made. The publisher remains neutral with regard to jurisdictional claims in published maps and institutional affiliations.

This Springer imprint is published by the registered company Springer Nature Switzerland AG
The registered company address is: Gewerbestrasse 11, 6330 Cham, Switzerland

Contents

About the Authors

Prabhakar Bandaru is Professor of Materials Science in the Department of Mechanical Engineering, University of California, San Diego.

Shreyam Natani is a graduate student in the Department of Mechanical Engineering, University of California, San Diego.

Chapter 1
Introduction

A most exciting development over the past decade is the discovery of new particles and related energy states in solid-state material systems. What was previously the province of exotic particle physics and massive accelerator systems can now be seemingly observed in tabletop experiments. For instance, moving beyond the electron (as the constituent of elementary charge) alternate fermion varieties such as of the Dirac mode or the Weyl mode have been observed in materials systems [1]. Such alternate modes and particles may bring forth new avenues of science and technology. Perhaps, just as electrons spurred *electronics*, it may just be possible that *Diracnics* or *Weylnics* may abound in the future, with hitherto unconceived of applications.

One such prospective application invokes quantum information science (QIS). Pertinent to QIS, associated new energy states, with zero or negligible interaction with the environment, are being considered for safely storing quantum information. Such aspects together with the quest toward new models of computation and numerical modeling have implications both for scientific curiosity and eventual technological application. In this context, new paradigms involving the harness of quantum mechanical states for information storage and transmission [2] has been considered to be very attractive. Much inspired by the notion that "*Nature isn't classical, dammit, and if you want to make a simulation of nature, you'd better make it quantum mechanical*" [3], there has been considerable effort to provide the hardware for the manipulation of the states that could, in principle, enable processing for a quantum mechanical simulation of natural phenomena—as pertinent to physical processes relevant in materials systems. For an engineered hardware, such basis states should.

(a) Be well defined and accessible.
(b) Inherently possess system relevant information.
(c) Be generally independent, while interacting when induced to do so through energy gradients/forces.

© The Author(s), under exclusive license to Springer Nature Switzerland AG 2022

P. Bandaru, S. Natani, *Topological States for New Modes of Information Storage and Transfer*, https://doi.org/10.1007/978-3-030-93340-1_1

However, such states would be expected to be intrinsically fragile as interaction with the environment seems inevitable even in the absence of any obvious external forces. Consequently, the criteria outlined above are difficult to fulfill in any physical system. While a *proper* interaction would indeed be a genuine simulation of natural phenomena, the dissipation of the information of the states through improper interactions (*noise* and *decoherence*) needs to be recognized and corrected, that is, through clearly designed *error correction* protocols. Such additional aspects for proper interactions necessary for quantum simulations implies a necessary overhead in the hardware, while not capable of clearly ensuring that all the undesired interactions have been accounted for and eliminated. Indeed, the necessity for a large number of properly configured states (*qubits*) continues to be a major hurdle in large-scale quantum hardware to enable the solution of problems that truly cannot be solved by existing *classical* computers in a reasonable time [4].

It is in the context of fulfilling the criteria for the quantum states amenable for qubits, that certain unique modes, classified broadly as topological states (TS), may be superior. A major consideration here is the *topological* aspect whereby such modes may be related to the configuration of a specific material. The identification of the state with the entire material naturally leads to a situation where the interaction of the particles within the material is irrelevant. Given that any well-understood particle would be interacting with its environment, for example, as a quasiparticle [5] in a material, such a noninteracting TS could alternatively arise through a collective excitation of the particles in the system. Here, the intra-particle interactions would significantly overwhelm the external interactions. Then, the TS could be considered immune to the environment to a significant degree. Of course, the extent of immunity must be clearly quantified, and such an aspect poses a problem from an experimental point of view, as will be discussed later. Assuming for now, that such an environmentally immune state is indeed feasible, and no further interactions are possible, it may then be inferred that the criteria related to (a) and (b) are fulfilled. Additionally, as the collective excitation have been invoked in the formation of the TS, the intrinsic/stored information is nonlocal with the implication that any local disturbance is inadequate to perturb the TS.

Other criteria for defining an energy level, corresponding to say a TS may be that at the very outset, the defined state should be immune to thermal energy–related disturbances. From the principles of thermodynamics, the temperature (T) is always finite and > 0, and a unique energy level corresponding to the TS may only be defined within an energy range of the order of k_BT, where k_B is the Boltzmann constant (~86 μeV/K). It may then be supposed that a lower temperature enables a more precise definition of the TS energy level. With such a consideration, we will now describe the possibility of a TS.

Given that a TS should be (i) nonlocal, and with a (ii) sharply defined energy level, one approach to experimentally harness such a state could be through the probing of lower-dimensional materials. The rationale is that the *hiding* of a TS in one of the dimensions, that is, the three space dimensions + one time dimension, would make the state immune to disturbances in the other dimensions. For example, a quantum dot or a nanowire, defined in zero- and one-dimension, respectively, could be probed for such a TS. Taken together with a clearly defined and distinct energy level, a suitable description of the TS would be in energy space, and commonly referred to as reciprocal/*k-vector* space. The reciprocal attribute arises from the de Broglie relationship connecting the momentum (p) and the wavelength (λ), through $p = h/\lambda$, where h is the Planck constant [6]. With k (= $2\pi/\lambda$), we see that the p is proportional to the k, and the energy–momentum relationship is mapped to an energy (E) – k-vector space. An alternative formulation involves the reduced Planck constant ($\hbar$) = $h/2\pi$, from which we obtain: $p = \hbar k$. It is noted that the *k-vector* depicts the direction of travel of the wave. A line of travel (in real space) is specified through a unique point in *k-vector/k*-space, while a surface in real space would be represented by a connection of such points, that is, a line in k-space.

Consequently, a spatially delocalized state (corresponding to posited formation of the TS from a collective excitation) with a specific energy would be represented as a point. The search for a suitable TS would then involve the search for a point suitably located in the E – k space that is not linked to other k-points. The related TS would be protected against environmental disturbances in that the no energy levels or bands should be passing through such a point.

As is well known, the mutual interactions between energy levels/bands, formed from the constituent atom–related electronic distribution in solids yield hybridized states, the interaction of which may be considered to be somewhat decoupled from those of the parent energy states [7, 8]. Such decoupling is obviously present, for instance through considering the bands of energy levels relevant to the two-dimensional surface [9]. The interactions of the surface states would be different than that of the bulk states [10]. Moving down dimensionality, we may consider lines on the surface and points on the line. The related energy states could be located in specific regions of the E-k space and may be considered distinct from bulk states. Intrinsic to such assumption is that the bulk may not be assembled from points, lines, and surfaces alone.

Given the distinct characteristics of lower-dimensional materials and the possibility of finding a TS in them, one is quickly led to the aspect that either ends (zero dimensional in nature) of the one-dimensional edges on a surface would be a likely candidate for probing greatly protected regions of E-k space. Such unique edge states may be manifested, for instance, in the quantum Hall effect [11] through the application of a magnetic field. The edge states bridge two bands with a slope corresponding to the velocity of carrier (electrons/holes) propagation along the edges. However, the external magnetic field may provide an undesirable external disturbance which detracts from obtaining an environmentally immune TS.

In this context, it was found that in specific material systems, broadly termed as topological materials, such edge states may be obtained through a magnetic field naturally intrinsic to the material, for example, through a spin–orbit (S–O) interaction [12–14]. Here, the intrinsic field arises from the interaction of the electron spins orbiting around a nucleus. The related magnitude of the field is expected to be proportional to the number of electrons and leads to the conclusion that such a S–O effect is more significant in higher atomic number elements, such as Bi, Hg, etc.

Indeed, the first investigations into topological materials involved systems incorporating compounds such as Bi_2Se_3, HgTe, and TaAs, for instance [15–17].

An alternate way of producing internal magnetic fields involves the consideration of curvature as a motivation. We are familiar with the notion that electrical currents moving in a circle may induce magnetic fields. However, the circle may be in real space or reciprocal space. In the latter case, the circle is constituted from a collection of k-points, each of which represents a direction, and there is hence a link between electrical carrier trajectory in real and reciprocal space. The edge states may then also be created through the connection of two *topologically distinct* materials, involving the merger of two energy bands. It is noted that both the potential energy as well as the kinetic energy ($= \frac{1}{2} mv^2 = p^2/2\ m = \hbar^2k^2/2\ m$) should be considered. While the former sets a reference energy level, the E-k dispersion nominally follows a k^2 variation for a free particle:

$$\mathrm{E} = \frac{\hbar^2 k^2}{2m} \tag{1.1}$$

An alternate E−k dispersion may be related to a linear k variation, as may occur for a light wave, propagating with a velocity (v), as in the following:

$$\mathrm{E} = \hbar v k \tag{1.2}$$

In the above relation, the velocity (v) may refer to a group velocity, say of the constituent electron waves at a given k. The *key* consideration is then that the overlap between bands must be reckoned in terms of the overlap of the related paraboloids (/cones) arising from the parabolic (/linear) variation. In the former case, the locus of points of the intersection of the paraboloids would be a circle of k-points. The carrier transport along the circle of points would result in an intrinsic magnetic field and related edge states.

As indicated previously, an *ideal* location for a state immune to environmental disturbance would be a single k-point, corresponding to a zero-dimensional situation. This implies a line in real space. A line could be defined on a surface of the material with the criterion that it is not connected to the bulk. The emergence of a related conducting edge *state*, at a very low interatomic spacing is indicated through Fig. 1.1, which also illustrates the closing and opening of the energy gap. The initial closure is due to the overlap of the wave functions between adjacent cells. With further reduction in the a, the surface states begin to be occupied as well and form conducting edges/*edge states*. While these states are derived from the bulk, they manifest conducting attributes distinct from the bulk states. The essential idea is then that conducting edge states would exist in the band gap of the bulk. Moreover, both these states exist in pairs, corresponding to the two edges. A *topological* attribute is evident in that such paired states are (a) formed at distinct spatial regions, and (b) located exactly mid-gap and are inaccessible/do not mix with the bulk states.

The above example is illustrative of the formation of paired edge states as formed through the intersection and subsequent separation of the overlap of the energy

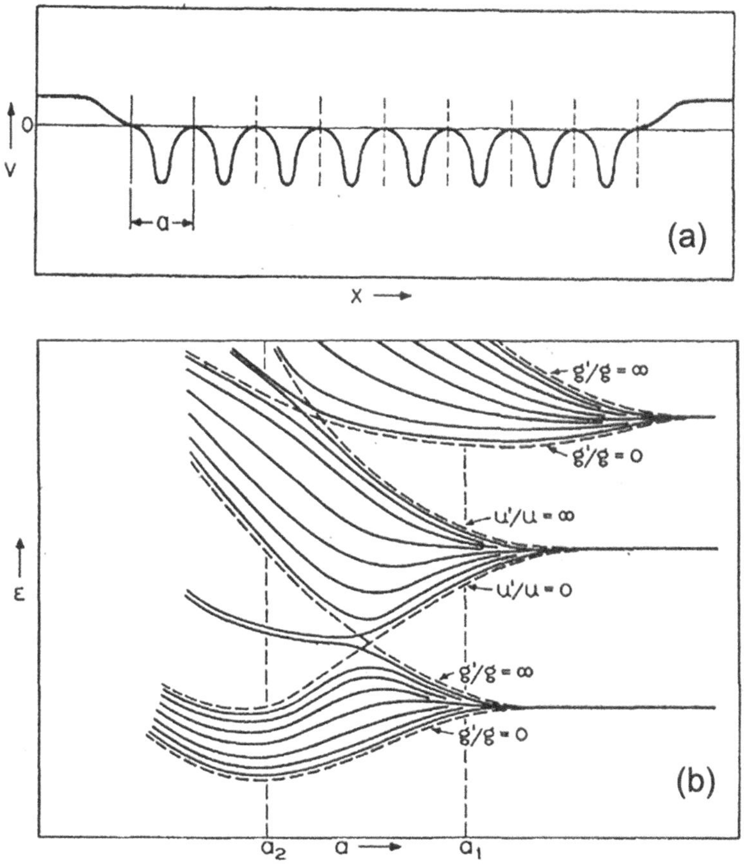

Fig. 1.1 (**a**) The periodic variation of the potential energy along a line of atoms, corresponding to a line on a surface. The *a* represents the interatomic spacing and the larger potential at the surface (left and right ends) is indicated. (**b**) The variation of the energy as a function of the *a* indicates a merger of the energy levels with decreasing *a* followed by an energy gap closure and a subsequent reopening of the gap. The emergence of two distinct surface states in the gap is noted. The surface states are half-occupied and hence metallic in character. Such states may be considered as *edge states* corresponding to the two edges of the one-dimensional chain. (Adapted from Shockley [9])

bands. The physical picture comprises three ingredients, for example, as envisaged through (i) the overlap of top and bottom paraboloids; (ii) the hybridization and change of character of the bands at the overlap/crossing; and (iii) an internal force causing the subsequent separation of the paraboloids, leaving behind a connecting *edge state.* The internal force may be provided by the S–O interaction related magnetic field, for instance. The criteria (i) and (ii) indicate metallic character, while (iii) necessitates that the specific constituent of the band be considered. For example, *nominally* an electron *d*-orbital constituted band may be expected to be at a higher energy level compared to a *p*-orbital constituted band. However, in a *topological* material, the *p*-orbital constituted band could be at a higher energy level due to material intrinsic fields such as from the S–O interaction [18], which interacts more with the bands related to the former. Now, at the interface of the *trivial* (where the nominal energy ordering is prevalent) and the *topological* materials, the mutual connection of the *p*- and *d*- derived bands necessitates the formation of an edge state, and is considered through a *bulk–edge* correspondence, where the edge characteristic bridges the two bulk attributes. Such connection, for example, termed a Berry connection [19], is more deeply understood within the spirit of gauge theory, with the associated phase of the energy- or charge-carrying wave. Such connection may also be visualized in terms of a bandgap closing and reopening in the transition from a *trivial* material to a *topological material.*

While the formation of conducting edge states at the ends, as indicated in Fig. 1.1, is then plausible, it is unclear whether these states may truly be considered as protected. For instance, the presence of another charge perturbation, say due to adsorbates, may influence the surface state and perturb the related energy levels in the gap, either upwards or downwards. Such perturbation would not however be expected in the case of a superconductor (SC) with particle–hole symmetry. Here, both positive or negative charges would simultaneously be involved in the perturbation, with the consequence that the mid-gap energy level is simultaneously moved down and up, that is, the energy level is stable. Such a symmetry arises from the consideration of Cooper pairs of electrons in the SC, where a pair is constituted from two distinct electron spins with net zero momentum, as represented by the pair ($\boldsymbol{k}\uparrow$, $-\boldsymbol{k}\downarrow$) [20]. The $\boldsymbol{k}$ and the -$\boldsymbol{k}$ may also be interpreted in terms of the characteristics of an electron and a hole [21]. At this juncture, we have posited that charge-immune states may be found in a SC and such states may possibly harbor TS. However, we have not yet considered the spin character of the electrons and there still exists a possibility of the flip of the spin state from say, up (/down) to down (/up). Such flips are considered possible as the electrons in the pair are close to each other in a typical SC. The proximity aspect arises from the Pauli principle relating to the antisymmetry of the overall electron pair wavefunction, constituted from the spin and the space components. Here, the antisymmetry of the spin component of the Cooper pair (from the up and down spins) should be associated with the symmetry of the space component, that is, proximity of the electrons. Indeed, most SCs are of such as *s*-wave type. However, for immunity to such flips, it may be beneficial to probe alternate SC varieties less sensitive to the spin, that is, *p*-wave SCs, with the attribute of a symmetric spin component and an antisymmetric space component. Here, the carriers involved in the SC are constituted from the same spin type, that is, either up *or* down, and are substantially separated. It is then expected that the corresponding edge states are protected and expected to be true TS.

In summary, environmentally protected states have been posited to occur in the lowest possible dimension, that is, as zero-dimensional edge states, which may be found, for instance at the edges of one-dimensional nanowires. There is a single associated *k*-state, and the possibility of isolation of such a state in energy. Given two edges in a wire, the energy state of the carrier is doubly degenerate. The immunity of the carrier to external charge perturbation is facilitated using a SC and further immunity to spin-related perturbations is done through use of a specific SC variety. Then, the two edge states (with the same energy) may be considered equivalent to each other, in terms of the same mass and spin but with opposite charge (corresponding to the inherent electron–hole symmetry) and termed as *anti-particles*. Such anti-particles may be related to environmentally robust TS and have been termed as *Majorana modes*. We will next seek to understand the underlying rationale for such mode formation together with the attempts to identify such modes in experiments.

1.1 Seeking Majorana Particles and Modes in Superconducting Materials Systems

We now consider how protected TS may be formed at the ends of an edge, considering the interactions of electrons in a collective system, such as a one-dimensional wire, in a solid-state material. For a complete description of the interactions from the viewpoint of presently known physics, the relativistic Dirac equation (incorporating the spin degree of freedom) is used in concert with notions related to quantum field theory (QFT) [22]. In one form, the Dirac equation is written as follows:

$$\left(i\gamma^{\mu}\partial_{\mu} - m\right)\psi = 0 \tag{1.3}$$

Here, the three components of space and one component of time is unified into the field (ψ) and the mathematical description demands 4×4 matrices (γ^{μ}). Typically, in the Dirac equation, the components of the matrices have both real and imaginary entries corresponding to a distinction between the space and time components, respectively. However, if all the entries are purely imaginary, both the particle and its reversed partner (defined as an *antiparticle*) would be identical. Such a situation was theoretically analyzed by Majorana [23], and it was deemed that the Dirac equation could still be valid with respect to the description of a spin ½ particle serving as its own antiparticle. While the electron and the hole could be considered as antiparticles, through such a point of view, in view of their opposite charge, an individual electron and hole by itself could not serve as their own antiparticle – per the Majorana prescription. Moreover, while a superposition of the electron and hole could be considered in terms of a composite particle, as in an exciton [24] ubiquitous in semiconductors, the net spin is unity which goes against a fermionic characteristic. Moreover, the electron and the hole masses may not be identical. Instead, a more harmonious superposition of the electron and the hole can be observed in the SC state. For instance, the Bardeen–Cooper–Schrieffer (BCS) theory of superconductivity [20, 25] invokes the formation of Cooper pairs involving the attractive interaction of electrons mediated through phonons. The electrons involved in the Cooper pair formation are of opposite spin and may be considered through a superposition mediated by the phonon interaction. The SC gap between the lower-energy electron states and the higher-energy hole states enhances the stability and rigidity of the SC phase. Such superpositions while not strictly constituting particles may be corresponded to Bogoliubov quasiparticles [20]. Moreover, such analogies to excitations and superpositions encourage the possible seeking of such Majorana particles in solids.

1.2 Mathematical Description of the Majorana Fermion Modes

The wavefunction of an electron may be distinguished from its complex conjugate. However, as indicated in the previous section, there is a possibility that both the wavefunction of a particle and its conjugate are equivalent, as in a Majorana-based description. The modes, which are now purely real and which satisfy such a description may be termed as Majorana modes. For a connection to reality, these modes must be interpreted in terms of the presence or absence of particles such as electrons. There would then be two MF modes, represented by "1" and "2" for a given electron site. We define $\gamma_{i,1}$ and $\gamma_{i,2}$ as corresponding to the real Majorana fermionic (MF) operators [26], as follows:

$$\gamma_{i,1} = c_k^+ + c_k \tag{1.4}$$

$$\gamma_{i,2} = \mathbf{i}\left(c_k^+ - c_k\right) \tag{1.5}$$

We have used the second quantization formalism, where particles such as electrons may be represented through a set of operators [27]. A creation operator (c_k^+) represents the notion of creating (/destroying) an electron (/hole) while a destruction operator (c_k) indicates the annihilation (/creation) of an electron (/hole). The + and – signs in Eq. (1.4) and Eq. (1.5) may be interpreted as relating to the in-phase and out-of-phase superposition of the related modes constituting the electron presence and absence. The i in 2(b) implies an orthogonality of the $\gamma_{i,1}$ and $\gamma_{i,2}$. Equivalently,

$$c_k = \frac{1}{2}\left(\gamma_{i,1} + \mathbf{i}\gamma_{i,2}\right) \tag{1.6}$$

$$c_k^+ = \frac{1}{2}\left(\gamma_{i,1} - \mathbf{i}\gamma_{i,2}\right) \tag{1.7}$$

Such an alternative interpretation indicates that an electron site, say at k is constituted from two Majorana modes. It should be noted that this is per se not equivalent to splitting an electron (!) but is instead an interpretation of whether the absence (/ presence) of an electron at a site may be understood mathematically in terms of local orthogonal spatial coordinates represented through $\gamma_{i,1}$ and $\gamma_{i,2}$. Since such coordinates are associated with a site they are protected. Moreover, the related coordinate axes are both real, that is, $\gamma_i^+ = \gamma_i$, just as x and y axes in a two-dimensional coordinate plane may *individually* be considered real. The **i** factor arises only when considering the superposition of the two axes to describe a point. In terms of an operator-based definition, the γ_i may be interpreted as Hermitian considering that the related expectation values are all real [27]. Considering a one-dimensional array of electrons, and interactions between electrons on the ith and the $(i + 1)^{\text{th}}$ site, there would exist a possibility of both *on-site* interactions ($\gamma_{i,1} - \gamma_{i,2}$) in addition to *inter-site* interactions ($\gamma_{i,2} - \gamma_{i+1,1}$). There is then a supposition that the constituents of the Majorana modes could be *isolated*, that is, through $\gamma_1$and γ_2.

The product $c_k^+ c_k$, is termed a number operator and counts the number of electronic states in a system. Such a second quantization based notation is well known and indicated in introductory quantum mechanics books [5, 27, 28]. For instance, the number operator, for example, when acting on a single-particle state yields a factor of one (from the initial reset of a state to zero and subsequent creation), and the sum of all such products could be related to the total number of states in the system. Just as the electron count could be considered through the number operator, we could then define a related Majorana number operator (n_i^{MF}) = $\gamma_i^+ \gamma_i$. However, as previously considered, $\gamma_i^+ = \gamma_i$, from the (a) related real axes, as well as (b) from the Hermitian characteristic of the operators, implying that $n_i^{MF} = 1$. Further, $\gamma_i \gamma_i^+ =$ 1, as well with the implication that (n_i^{MF}) = $\gamma_i^+ \gamma_i$ would now be zero, from the anti-commutation relation: $\{ \gamma_i, \gamma_i^+ \} = 1$. Note that $\{ \gamma_i, \gamma_i^+ \} = \gamma_i \gamma_i^+ + \gamma_i^+ \gamma_i$. Intriguingly, the n_i^{MF} would now *both* zero and one! The counting of the number of Majorana modes then "*does not make any sense*"[29] and its presence would then be manifested *only* from the background ensemble.

For instance, a SC, in which Majorana modes may be manifested, may consist of an even or an odd number of electrons, defined in terms of an even or odd parity, and an equivalent representation in terms of a |0 > or |1 > state, respectively. Then, the induction and presence of a Majorana mode may be detected through the change of parity, that is, a change of state from |0 > to |1 > or vice versa. For instance, the addition (/subtraction) of an electron from an SC would modulate the $\gamma_{i,1}$ and $\gamma_{i,2}$, as inferred from Eq. (1.6) and Eq. (1.7). Here, c_k|1 > = |0>, c_k^+ |0 > = |1>. Note that each state is defined by two MF operators. Consequently, two electron operators and correspondingly four Majorana modes are the *minimum* number necessary to effect a given quantum state consisting of a superposition of |0 > and |1>, as in a *conventional* qubit of the form: a |0 > + b |1>. Here, a and b subject to normalization constraints, and where the entanglement yields additional degrees of freedom [2]. A MF mode–based qubit, based on an additional topological degree of freedom (based on the *inter-* and *intra-*interactions of electrons or equivalently the MF modes), could be the basis of operating on and/or modifying a physical state. A particular system where the even or odd characteristic is clearly manifested is a superconductor (SC)—which contains nominally electron pairs (corresponding to even parity) as well as a normal component (corresponding to single electrons and odd parity). It has been considered plausible that information could be stored in a Majorana mode, which is *by definition* enabled through considering the collective state of the delocalized ensemble, as in a bulk SC.

1.3 Majorana Modes for Quantum Information Science (QIS)

If information could be stored in the MF mode, the robustness to local perturbations, inherent to a specific/intrinsic definition of the respective coordinates, implies a high degree of immunity to local environment–related decoherence. However, operations involving MFs may yet be accomplished, for example, exchanging one MF with another. Such an exchange aspect lies at the core of the definition of a particle. For instance, basic quantum mechanics defines a particle through a wavefunction description, for example, through the sign of the wavefunction, which is unchanged (/changed) when the positions of two bosons (/fermions) are exchanged. However, for non-Abelian particles such as the MFs, a particle exchange, say by looping one MF around another MF, does not yield just a sign change but instead an arbitrary change of phase to *any* value. The *anyon* character of the MFs [30] is at the heart of the additional topological degree of freedom. In this context, a braid operator B_{12} is defined through.

$$B_{12} = \frac{1}{\sqrt{2}}\left(1 + \gamma_1\gamma_2\right) \tag{1.8}$$

For a particular Majorana mode, $\gamma_i = B_{12}\gamma_i B_{12}^{+}$, whereby $\gamma_1 \rightarrow -\gamma_2$ and $\gamma_2 \rightarrow \gamma_1$. The implication is that an interchange of the Majorana states is possible through the action of the braid operator. It would then be plausible that an effective particle-like configuration, and related operations, may be achieved through such *braid-like* operations.

Given the fragility of the quantum states to unwanted interactions, see Sect. 1.3, there are stringent requirements on the proper definition of the constituent basis for the states, for example, a clearly defined up or down spin, clockwise or counterclockwise polarization, etc. In a similar spirit, the Majorana modes which form the basis for the even–odd parity $|0>$ or $|1>$ state, respectively, should be considered within the purview of the commonly accepted DiVincenzo criteria [31], for their suitability in quantum information storage (QIS and subsequently quantum computation). We will now enumerate the relevant notions along with the extent to which the Majorana modes can be satisfactory.

1. The Hilbert space relevant to a quantum system should be clearly defined and *grow* with the number of particles in an exponential manner. Such an increase is related to the scaling of the quantum system with increasing degree of entanglement. The Majorana modes may extend over a multidimensional and the space for performing the computations would increase with the number of modes.

Given that an ensemble state yields two Majorana modes, for example, corresponding to the two ends of a nanowire, in principle N nanowires would be expected to yield 2^N dimensionality of the Hilbert space.

2. It is then essential to initialize the quantum system to a well-defined *fiducial* state, that is, all the Majorana modes should be clearly set to the same basis. As the Majorana can be defined through the "zero-energy" mode description, the related criterion is fulfilled.
3. The aspect that the basis states should be well isolated from coupling to the environment was considered to be a "*tough requirement*" [31] generally. It is here that the Majorana mode basis is particularly useful as the modes themselves are not physical, referring to the local coordinate definition of a particle such as an electron, and hence are not subject to physical constraints from the environment! Instead, the modes encode the state of the system viewed as a collective ensemble and are not perturbed unless the ensemble changes which implies another system! For instance, the conversion of a $|0>$ to a $|1>$ state, one hallmark of a decoherence is explicitly forbidden through the use of MFs. The rationale for such *exceptional stability* is that there is no particular ground state to which the MFs can be stabilized.
4. For practical utility, say in processing quantum information, it should be able to modulate the basis states of the quantum system through unitary operations. The braid operator (B_{12}) is equivalent to a rotation of the qubit state through an angle of 45° corresponds to such an operation. Moreover, unitary operations which may introduce bit flip errors in a conventional qubit may be reduced through deploying braiding which is sensitive to the orientation or topology of the braiding and not particularly to the local dynamics or interactions of the MF.

When the braid operator acts on any one particular state, that is, $|0>$ or $|1>$, it causes a change of phase through $B_{12}\,|0> = \frac{1}{\sqrt{2}}(1+i)\,|0>$, or $B_{12}\,|0> = \frac{1}{\sqrt{2}}(1+i)\,|0>$. The implication is that the braiding or exchange of Majorana states, *belonging* to one electron, with another is only manifested through a change of phase. To perform a unitary operation, which modulates the phase as well as the relative magnitude of the constituent ensemble states, the use of MFs from two different electrons would be necessary. Given a collective state of the two electron system, say $|00>$, with four related MFs (γ_1, γ_2) and (γ_3, γ_4),both $B_{12}\,|00>$ and $B_{34}\,|00>$ would only induce a phase change while $B_{23}\,|00>$ would yield a superposition related to different number states, for example, $B_{12}\,|00> = \frac{1}{\sqrt{2}}(1+i)\,|00>$, while $B_{23}\,|00> = \frac{1}{\sqrt{2}}(|00> + i\,|11>)$. In the latter case, the number parity of both the $|00>$ and the $|11>$ state are relevant.

5. The quantum system should then be sufficiently robust that *strong* measurements, accessing the inherent basis states, could be accomplished. Such measurements should be done within the lifetime/decoherence time of the Majorana modes, which again due to their immunity from environmental disturbance would be possible.

1.4 Experimental Realization of the Majorana Modes in Superconductor Materials

It is well known that in a SC that the electrons may all be considered as paired (as Cooper pairs with oppositely oriented spins) and consequently an added electron would be related with a corresponding Cooper pair hole. The superposition of an electron and hole in the SC, corresponding to the added/odd electron, may then be related in a single-particle-like picture to the formalism related to Eqs. (1.6) and (1.7) with related Majorana modes, with an associated collective state arising from the Cooper pair background dominated ensemble.

An interesting aspect, given the single added electron as well as the *half-filled* aspect of the Cooper pair, is related to the a priori indeterminate electron spin (it could either be up or down spin). Such a situation is akin to ignoring the electron spin altogether. As one is dispensing with the notion of electron spin, specific SC materials where this is possible must be considered for hosting the MFs. Moreover, a strict symmetry between the electron and the hole state is implicit in the creation of Majorana states/modes. Consequently, energy structure symmetry in the electron and hole states is necessary, with the implication that details of the band structure *should* only play a minor role.

A SC is a natural system where such symmetry is observed, and the specific band structure aspects are not at the forefront, in expository treatments of SCs [20, 31]. However, the prototype s-wave SCs are, at the very outset, an unlikely candidate for hosting the MFs, as electron pairing is in the singlet configuration, and hence the spin character, *is* quite important. The Pauli principle [33]—with the premise of an overall antisymmetric total eigenfunction for the electron system (where either the spin part *or* the space part of the eigenfunction is symmetric or antisymmetric) is invoked. In an s-wave SC, the paired electrons (i.e., symmetric space part) exist in a spin singlet configuration, that is, antiparallel spins yielding an antisymmetric spin component to the wavefunction. However, when the electrons are in a triplet configuration, that is, parallel spins either both oriented up or down with symmetric spin part, the spin character is not explicit. The corollary is an antisymmetric space constituent of the eigenfunction, that is, related to delocalization. Such an aspect is evident for electrons involved in a p-wave or even a d-wave SC, corresponding to a finite angular momentum and a delocalized electron system yielding an effectively *spinless* system. It was on such a premise that Kitaev was able to build a toy model of a 1D p-wave SC tight-binding chain [34]. Here, one considers the interactions *at* an electron site and *between* mutual sites. As may be inferred from Eqs. (1.6) and (1.7), and from the notion that there is one additional electron (a SC with odd parity), each electron site is constituted from the two MF operators. Considering a situation where the inter-site coupling is stronger compared to the intra-site coupling, a situation corresponding to the formation of the MFs, at the ends of the wire, as well as pertaining to an additional electron in the system, may be envisaged. Alternately, the engineering of a transition between a phase corresponding to paired electrons (a trivial phase) and a phase with MFs at the ends (a topological phase) can be invoked as indicated in Figs. 1.2 and 1.3.

A major issue, even at this stage, is whether the electrons in a *p*-orbital would truly be considered to be spinless as opposite orientation in the orbital would be expected from Pauli exclusion principle and the Hund rules of basic quantum chemistry [35]. Indeed, such a Fermion doubling–related problem needs to be overruled by the tacit assumption that half of the degrees of freedom of the electrons constituting the chain in the *p*-wave SC are frozen out [36]. If such a situation could be established, then the formed Majorana modes, associated with the additional electron added to the SC, would be situated mid-energy level between the ground state and the one-electron excited state in the SC and be construed as zero-energy modes, providing a basis for their initialization and reinterpretation in terms of the Majorana modes.

1.5 The Environmental Robustness of Majorana Modes

Considering the equal superposition of the electron and the hole operator states (in the half-filled Cooper pair) corresponding to a Majorana mode: Eq. (1.4), the relevant mode is stated to have zero energy. With respect to an SC gap (Δ_{SC}), corresponding to the energy separation between the SC ground state and the first excited state, the mode would strictly be at mid-gap. The delocalized aspect of the modes in real space corresponds to a specific *k*-point in the energy band structure. Such delocalization in real space and the isolated aspect in energy space affords a prime example of a topological state (TS). The unique aspect of a MF seems to be that it arises from a collective state and is hence properly classified as a mode of the *p*-wave SC system. Consequently, the identification of the MF would hinge on

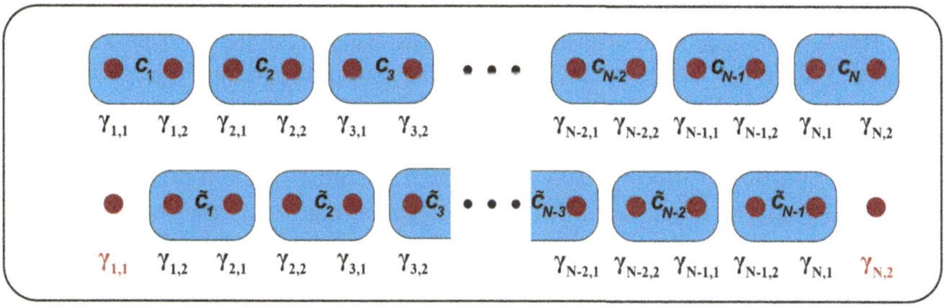

Fig. 1.2 A version of the original Kitaev's 1D p-wave SC tight-binding toy model explaining the formation of Majorana modes in condensed matter systems. Upper panel: Each fermionic operator c_i is split into two Majorana operators $\gamma_{i,1}$ and $\gamma_{i,2}$. The red dots denote the Majorana fermions and the blue box represents an electron. Lower panel: In the limit of $\mu = 0$ and $t = \Delta$, the fermions can be created by combining two Majorana fermions from neighboring sites ($\gamma_{i+1,1}$ and $\gamma_{i,2}$). This leaves two unpaired Majorana fermions at the two ends of the 1D system. It is important to notice that both cases have the same bulk properties

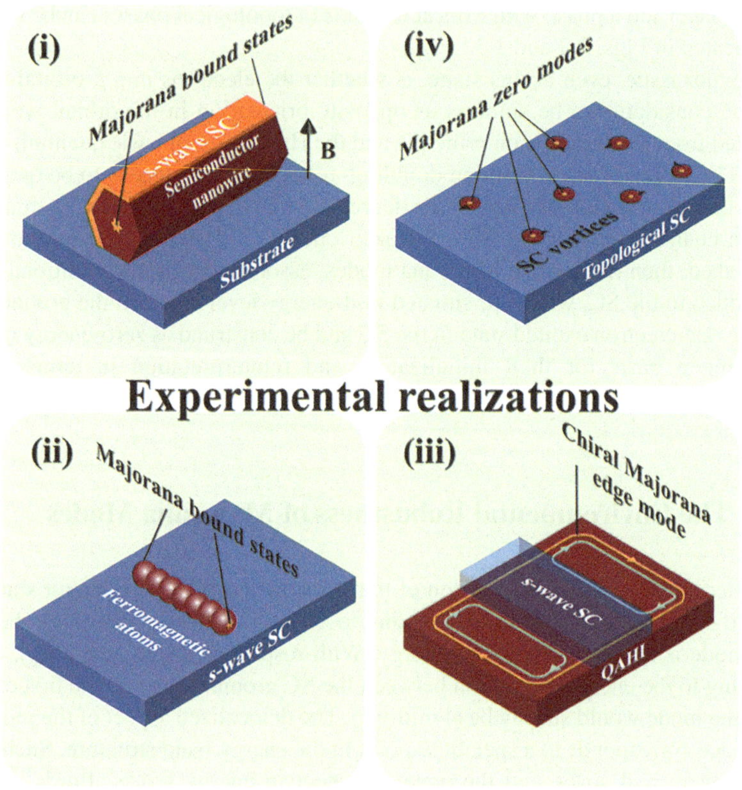

Fig. 1.3 Majorana modes and their manifestation. A summary of practical materials systems where such MF modes have been investigated. *Lower left panel*: A ferromagnetic atom chain placed on a s-wave SC with strong spin–orbit (S–O) coupling induces two Majorana bound states (MBS) at the two ends of the chain. *Upper left panel*: A semiconductor nanowire with strong S–O coupling partially covered by a s-wave SC fosters two MBS at the two ends of the nanowire. *Lower right panel*: A *s*-wave SC placed on top of a Quantum Anomalous Hall Insulator (QAHI) results in chiral 1D Majorana edge modes which travel along the boundary of the SC. *Upper right panel*: Abrikosov vortices in a topological SC (each allowing a magnetic quantum flux to pass through) are thought to contain Majorana zero modes at the center of the vortex

coupling to particular mode in a system. The robustness of the topological mode of the MF may also be considered through the aspect that unless the overall system is changed, for example, through the addition of an electron to the odd parity SC, only then would the MF be destroyed. Such a change corresponds to a global variation and as a corollary all other changes pertaining to less than the addition of an electron, for example, electron scattering in the system, electron–nuclear interactions, etc. may be considered as local perturbations and would not change the underlying aspects related to the formation of the Majorana mode.

An alternate point of view, related to the robustness of the Majorana mode is in terms of its unique mid-gap energy state and any event that disturbs the mode to an extent of less than ½ Δ_{SC}, would not affect the mode and the information carried by the mode. Moreover, the electron–hole symmetry intrinsic to the mode implies scattering between energies in the SC gap embracing the Majorana state. However, there may yet be states symmetrically placed around the zero-energy mode between which scattering could yet occur. For instance, alternate in-gap states in the SC reduce the magnitude of the *protective* energy gap to below Δ_{SC}. A further decrease in protection could be through the interaction of the Majorana modes, which could occur for instance when the electron interactions are screened, would correspond to a form of *quasiparticle poisoning* and destroy the MFs. Indeed, such *poisoning* could indeed be a *major issue* in the formation as well as the realization of the MFs. Alternately, electrons tunnelling into a Majorana mode, based on presently known methodology for inquiring into such states, could also couple with the mode and lead to poisoning. Obviously, such quasiparticle-related effects may spoil the protected nature of the Majorana modes with major implications to utility for quantum information storage.

Chapter 2
Practical Materials Systems, and Related Criteria, for Hosting the Majorana Modes

However, the very notion of a *p*-wave SC is fraught with whether such materials even exist! Pertinent to this concern is an long-standing issue in the SC field is whether non *s*-wave SCs even exist, that is, whether there would always be a mixture of *s*- + *p*-wave character [37, 38]. For instance, in the high T_c SCs, an admixture of *s*- + *d*-wave character has been considered [39]. Nevertheless, even in seemingly well-established systems, there has always been a reconsideration of the *s*-wave characteristic [40]. Consequently, a breakthrough was achieved when it was considered that *s*-wave SCs in *proximity* with particular topological materials could induce unique states, such as the Majorana modes [41]. Consequently, isotopically oriented *s*-orbital-related energy levels would have to be considered for the formation of the modes.

In this context, it is well known that the large coherence lengths (the distance over which the superconductivity permeates, say into a normal material [8]) inherent to a SC material makes its presence felt through *inducing* superconductivity in an adjacent nominally non-SC material, with certain unique characteristics related to the delocalized states. The adjacent material could be comprised of a topological insulator (TI), an outstanding feature of which is that the expected energy levels are *inverted* with respect to what is nominally expected. For instance, the *s*-orbital-related energy is typically lower than the *p*-orbital energy, and accords with the expectation of a *trivial* insulator due to the related energy gap. However, in a TI, an energy-level inversion where the *p*-orbital-related energy is lower than the *s*-orbital energy is possible. Then, the *p*-orbital-related levels are involved in the proximity SC of the TI, and could facilitate the formation of Majorana modes.

Such an energy-level inversion occurs in TIs through an internal magnetic field arising from the spin–orbit (S–O) coupling-related interactions. The internal field may also cause the formation of edge states, as may appear in the ordinary or the anomalous quantum Hall effect (QHE) [11]. Such topologically protected states occur at the surface and may be coupled into through proximity with the SC. We now have a situation where a *s*-wave SC layer placed onto a TI induces

© The Author(s), under exclusive license to Springer Nature Switzerland AG 2022

P. Bandaru, S. Natani, *Topological States for New Modes of Information Storage and Transfer*, https://doi.org/10.1007/978-3-030-93340-1_2

superconductivity in the latter, arising from the tunnelling of the Cooper pairs from the SC to the surface states of the TI. Moreover, the related spin-momentum locking between the electrons from the SC and the edge states in the TI would yield an anisotropic/*p-wave* character to the Cooper pairs.

2.1 Topological Insulators (TI)—With Induced Topological Superconductivity (TSC)

The S–O coupling, relevant to TIs, is broadly relevant to compounds with heavy elements, such as InSb or Bi_2Te_3, and may be used to induce unique vector potentials and equivalent magnetic fields with distinct topological signatures [42]. The surface states of TIs may be thought of as having infinite spin–orbit coupling, as carrier propagation along such states is infinitely unidirectional. When one-dimensional (1D) nanowires (NWs) constituted from such compounds are coupled to SCs, proximity effects [41] induce SC concomitant states *in* the NW. Zero-energy states may be posited to exist at the point midway between the top and the bottom edges of the induced SC energy gap [43] in the TI, and may be further modulated through an externally applied magnetic field (B_{ext}). Considering then both the (a) 1D characteristic, and (b) S–O coupling–induced magnetic field polarity, the zero-energy states in the proximity-induced SC NW are pushed to the two ends, where they may be construed to be Majorana modes, as in Kitaev's proposal and related to Fig. 1.2. Indeed, a tendency toward modulating the mutual electronic interactions would be facilitated through *both* an internal magnetic field (B_{int}), say from the S–O coupling, as well as the B_{ext}. The mutual interplay of such fields would assist in the formation of energy states, for example, at the ends of NWs, as manifested from unidirectional edge modes.

We will first consider the influence of the spin–orbit (S–O) interaction, which has been considered responsible for the energy inversion, as related to lowered *p*-orbital-related energies. We will see that the S–O interaction induces a magnetic field (B_{SO}) which *pushes* up the *s*-energy states.

2.2 The Spin–Orbit (S–O) Interactions in a Nanowire (NW)

Considering a simplistic classical image of an electron orbiting around a nucleus yields an aspect related to a magnetic field, even at the atomic level, and may be manifested as an intrinsic magnetic field (B_{int}). From a Newtonian mechanics as well as relativistic electrodynamics perspective, the spin motion of electrons orbiting around a nucleus couples the related electrical fields ($\boldsymbol{E}$) with the electron velocity ($\boldsymbol{v}$), yielding an effective *intrinsic* magnetic field

$$\boldsymbol{B}_{\text{int}} = \left(\frac{1}{c^2}\right)(\boldsymbol{v} \times \boldsymbol{E}) \tag{2.1}$$

A related magnetic energy (the product of the force and the distance) is indicated through the product $\mu_B \cdot \boldsymbol{B}$ (where the magnetic moment of the electron [44]: $\mu_B = e\hbar/2m$). The $(\frac{1}{c^2})$ v factor arises from the related Lorentz transformation between coordinate frames [45], associated with a static nucleus and a rotating electron. It is interesting to compare the effects of the $\boldsymbol{B}_{int}$ with the traditional Hall effect, where the electron motion in a static magnetic field results in an orthogonal electric field, while in the spin–orbit case, an electron in a constant electric field is subject to an orthogonal spin–orbit related magnetic field.

We have seen that the interaction of the electron spin with the orbital precession of the electron around a central nucleus is involved. With a proper representation of an electron, considering the orientation, that is, spin, in terms of two-dimensional Pauli matrices ($\boldsymbol{\sigma}$) [46], the *spin–orbit* interaction energy (Δ_{SO}) would be as follows:

$$\Delta_{SO} = \mu_{\mathbf{B}} \cdot \mathbf{B} = \frac{e\hbar}{2mc^2} \sigma \cdot [\mathbf{v} \times \mathbf{E}] \tag{2.2}$$

While the magnitude of the E_{SO} could at first sight be extremely small, for example, due to the c^2 term, the $\boldsymbol{E}$ could be large due to the small distances involved. Consequently, the E_{SO} is estimated to be in the range of three orders of magnitude from ~6 meV (for light elements such as carbon) to ~1 eV (for heavy element constituted compounds such as HgTe). A larger atomic number is typically related to a larger $\boldsymbol{E}$ and explains the larger E_{SO} for heavy elements. Further, the E_{SO} acting on the spins of the *s*-orbitals of the proximally induced SC in the TI would push up the related energy levels. Consequently, with the lowest energy state constituted from the *p*-orbitals and mapped to the SC ground state, we obtain *p*-wave like/topological superconductivity. The essential ingredients for the manifestation of the Majorana modes are hence obtained.

2.3 Manifestation of Topological Superconductivity in One-Dimensional Systems

Topological superconductivity (TSC) would occur when the nominal SC interactions are reversed in a topological material, through energy modulations/reciprocal space–dependent interactions. This requires *both* the breaking of spin symmetry which is facilitated through internal/intrinsic magnetic field (say, due to the S–O coupling: B_{SO}) or extrinsic magnetic fields. Such an *intrinsic* magnetic field would be added to the B_{ext}, and related coupling could result in band structure alteration. The principles underlying the formation of a topological SC phase are indicated, through related band diagrams, in Fig. 2.1.

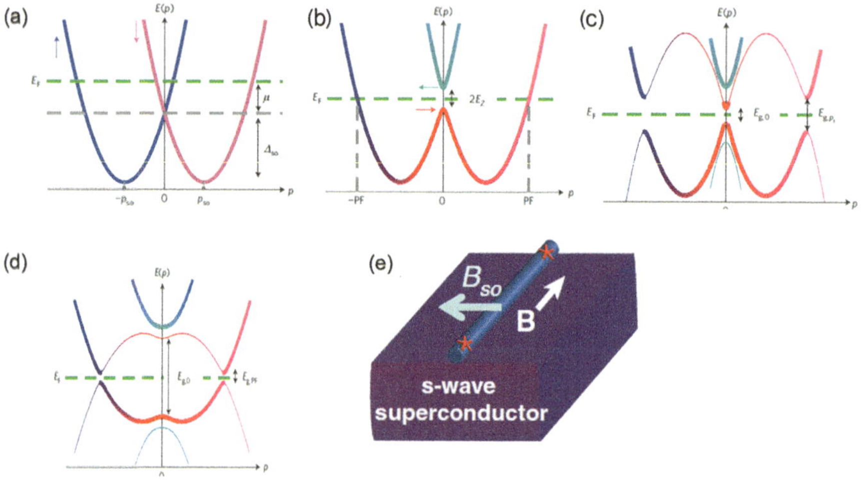

Fig. 2.1 *The evolution of the band structure in a topological superconductor for the provision of Majorana modes*. (Adapted from [50, 79]) (**a**) The energy–momentum (E–$\boldsymbol{p}$) space band structure for the two spins in a topological insulator (TI), resolved through a spin–orbit (S–O)-induced intrinsic magnetic field (B_{SO}). The resolution of the down (↓) and up (↑) spin states is shown. The external magnetic field, $B_{ext} = 0$. The chemical potential ($\boldsymbol{\mu}$) [47] corresponding to the Fermi energy (E_F) is indicated. (**b**) With $B_{ext} \neq 0$, and perpendicular to B_{SO}, the mixing of the two spin bands occurs, along with the opening of a Zeeman gap (of energy magnitude = $2E_z$) at the center of the Brillouin zone, the edge of which is at the Fermi momentum ($\boldsymbol{p}_F = \hbar \boldsymbol{k}_F$). The resolution of the spin states in the perpendicular direction, that is, → and the ←, due to the B_{ext} are shown. (**c**) The band structure with induced superconductivity and a related SC gap at the $\boldsymbol{k}_F$. The band structure from the TI hybridizes with that of the SC (electrons: bolder lines) as well as the SC (holes: lighter lines) are shown. (**d**) An increasing B_{ext} leads to a larger E_z which dominates over the trivial SC gap and leads to a *nontrivial/topological* SC. The band inversion between when $\boldsymbol{\Delta}_{SC} > E_z$, and when $E_z > \boldsymbol{\Delta}_{SC}$, proceeds through a formation of a net zero-energy state corresponding to the Majorana mode, (**e**) The possible location of the Majorana-bound states (red*) at the end of a NW with proximity superconductivity induced from the *s*-wave SC below. The B_{ext} (= **B**, in the figure is applied along the wire perpendicular to the spin–orbit-coupling-related magnetic field (B_{SO})

We first note that for a NW that the energy dispersion is governed by Eq. 1.1 along the long direction, along with energy quantization in the two perpendicular directions, that is, along the NW diameter. Then, with $p = mv = \hbar k$, the $\Delta_{SO} = \frac{e\hbar^2}{2m^2c^2}\sigma[\mathbf{k} \times \mathbf{E}]$, implying that for a given momentum ($\boldsymbol{p}$ or $\boldsymbol{k}$) along the NW corresponding to an electrical current, and an $\boldsymbol{E}$ perpendicular to the NW, there would be an effective spin dependent E_{SO}. The spin degeneracy is then broken and is manifested in the energy band diagram through two separate E-$\boldsymbol{k}$ dispersions, as indicated in Fig. 1.2. The shift of the origin from $\boldsymbol{k} = 0$, by $\pm \boldsymbol{k}_{SO}$ ($= \boldsymbol{p}_{SO}/\hbar$), is related to the $\pm\Delta_{SO}$, corresponding to the down (↓) and up (↑) spins. Each of these bands may be filled up to an energy level corresponding to the Fermi energy (E_F) or the corresponding electrochemical potential (μ) [47].

In summary, when $B_{ext} = 0$, the energy momentum dispersion corresponding to the up and down spins are displaced along the $\boldsymbol{k}$-axis corresponding to the symmetry relation: $E_\uparrow(k) = E_\downarrow(-k)$ due to the B_{SO}. When B_{ext} is applied along the NW, there would just be a skewing of one set of curves with respect to another, but when B_{ext} is oriented perpendicular to the NW, the two *spinful* bands are mixed and the zero-field crossing point is converted to an anti-crossing and a resultant S–O gap [48]: Fig. 2.1b. The resultant Zeeman field yields an energy band gap of magnitude twice the E_Z (= $g\mu_B B_{ext}$), with g as the Lande g-factor and μ_B as the Bohr magneton (~86 μeV/T). A related S–O energy gap at the $\boldsymbol{k} = 0$ point is manifested. Typically, the g-factor is large at ~50 in InSb, implying the use of smaller B_{ext}, so as to lessen the perturbation of the SC characteristics. Consequently, for a typical B_{ext} of ~0.1 T at 20 mK, the E_Z for InSb would be of the order of 0.2 μeV.

Subsequently, when the s-wave SC (with electron singlet states) is placed in proximity, superconductivity as defined by a related energy gap (Δ_{SC}) is induced in the NW and is manifested in the band structure at the edge of the BZ corresponding to the Fermi momentum/wave-vector (k_F): Fig. 2.1c. The SC energy gap (Δ_{SC}) is related to the critical temperature (T_c) through the relation: $\Delta_{SC} \sim 3.5\ k_B\ T_c$ in the BCS theory [20] and may not be identical to the gap in the induced SC. It should be noted that the singlet aspect induces a coupling between the up and down spins of opposite momentum. The coupling is facilitated through the μ being situated in the Zeeman gap. Generally, the smaller of the band gaps (at $\boldsymbol{k} = 0$ or at $\boldsymbol{k} = k_F$) determines the overall system characteristics. For instance, when there is a finite and $E_z = 0$, the state is that of a regular/*trivial* SC. However, when $E_z > \Delta_{SC}$, the regular SC properties are overwhelmed and correspond to a *topological/nontrivial SC*.

Generally, when μ is small and positive relative to the difference between E_z and Δ, that is, $\mu < \sqrt{E_z^2 - \Delta_{SC}^2}$, then the topological SC phase persists. The transition from trivial to nontrivial SC phases occurs when the combined influence of the μ and the Δ_{SC} are equivalent to the E_z, that is, at $\mu = \sqrt{E_z^2 - \Delta_{SC}^2}$. Here, the system passes through an effectively zero-energy state, where the *net* difference between the internal energy (as related to the μ and the SC gap) and the external energy, that is, from the opposing Zeeman-related energy band gap, is zero. The system state may now yield a Majorana mode [49].

Further, as B_{ext} is progressively increased, the $E_z > \sqrt{\mu^2 + \Delta_{SC}^2}$, and a band inversion related to a topological phase transition would occur. For instance, with increasing B_{ext}, the lower (/upper) bands determining the effective energy gap which were originally situated at $\boldsymbol{k} = 0$ are now located at $\boldsymbol{k_F}$: Fig. 2.1d. The main idea is that a B_{ext} would yield enough energy to push the lower electronic energy level past the intrinsic energy scales set from the electron concentration: μ, and from the SC: Δ_{SC}. The phase transition occurs at a single k-point and yields delocalization in real space along the length of the wire. As indicated in Fig. 2.1 and also verified experimentally [50] that only when B_{ext} is perpendicular to B_{SO}, for example, by rotating B_{ext} in a plane perpendicular to the substrate, is the conductance of the zero-energy state a constant. However, even when B_{ext} is rotated in the plane of the substrate there are certain angles where such maximal conductance was obtained.

The zero-crossing of the energy levels/bands and the subsequent inversion of the energy manifold has been indicated as an ingredient for the formation of topological phases, such as the Majorana modes. At this crossing point, as with increasing B_{ext}, the inter-electron interactions would dominate and lead to a delocalization of the modes. The maximum extent of delocalization corresponds to the modes to be at the ends, that is, of a NW/atomic chain, nodal line in a vortex, or an edge state – as depicted in Fig. 1.3. The formed topological phase is stable due to the utilization of *both* the external and internal fields, that is, with SC related to broken spin and particle–hole symmetry, and it is assumed that beyond the phase transition, the zero-energy modes would be delocalized.

2.4 Probing Majorana Modes: Issues Related to Finite System Size and the Possibility of Alternative Zero-Energy Modes

Given the delocalized nature of the Majorana state, it is unclear at the very outset, whether the related modes can be probed by local measurements. Both dynamic and static characterization techniques have been proposed to probe the energy states corresponding to the related modes. The passage of a current through such an energy state would be a clear signature of the existence. This involves tunnelling between equal energy states, which is in principle, possible, considering a normal metal contact–topological SC interface. Given the one-dimensional nature of the electrical conduction process, the conductance related to tunnelling between an electrical contact state and a MBS is expected to yield a peak value of the conductance quantum ($G_o = \frac{2e^2}{h}$). Alternately, this would correspond to a value of ~77.5 μS (the reciprocal of which is 12.9 kΩ). A related sharp increase in the current through the alignment of the energy level in the contact with the MBS state, would be manifested as a peak in experimental measurements. The voltage/energy width of the peak would be expected to be related to the electron temperature as well as the contact–NW barrier height.

While seemingly straightforward, such probing of the conductance and electrical current for the Majorana states, relies on the solitary presence of the corresponding *unique* MBS. However, we have seen that the MBS are formed in pairs, and ideally the distance between the pairs is infinitely large—corresponding to a precise energy and uniquely localized *k*-state and implying a very large delocalization length and an infinitely long NW. This is not practicable in an experimental system and hence the peak conductance could be modulated depending on the distance between the pairs of the MBS. Moreover, it is also being assumed that the topological superconductivity has been precisely tuned (as indicated in the previous section) and all that exists is the MBS. However, there are many unrelated phenomena that may also yield such effectively zero-energy states.

Andreev tunnelling–based phenomena are invoked when an electron from a normal metal electrode is incident on proximity to SC-bearing nanowire. The creation of a Cooper pair, in the SC, is then typical and is accompanied by the reflection/

backwards transmission of a hole in the normal metal [20], together with perfect resonant transmission through the SC. The process of Andreev tunnelling through a contact (i.e., normal metal) SC–contact (i.e., normal metal) sandwich structure also invokes the formation of electron–hole-bound states (Andreev-bound states: ABS) in the SC. An elementary exposition of the formation of such a state is indicated in Fig. 2.2.

Such discrete ABS exist in the SC gap and, indeed, the merging of the ABS at mid-gap/zero-energy was also posited to yield the Majorana modes [51] or more exactly Majorana-bound states (MBS)—as the mode is not independent of the underlying materials system and is bound to an underlying topological feature, such as a surface or the ends of a NW. Consequently, MBS may even be construed as a subset of the total set of ABS states related to the zero-energy manifold. However, true non-abelian MBS states—representative of states capable of being used for topological quantum computation would then be a subset of the total set of ABS states. The non-abelian MBS would need to be discriminated from trivial states through a careful interpretation of the relevant charge tunnelling processes [52].

There is also a possibility that MBS may not be forming at the ends of the wire, for example, when it is pinned by defects, and such an aspect confuses quite considerably their investigation and clear discrimination from the "*garden variety*"/"*standard*" ABS [52]. It was observed that even in the presence of a topological phase transition, the conductance map looks similar to those previously observed [51] utilizing only ABS-related overlap. Indeed, such ABS state coalescence was indicated also in the non-topological regime in the absence of obvious disorder. Hence, although a stable ZBP is a necessary condition for the formation of the MBS, it is not sufficient. Consequently, three classes of zero-energy modes were posited in semiconductor-SC-based hybrid systems [52]:

1. Standard ABS-related modes, which may correspond to strongly overlapping MBS. The overlap range is smaller than a characteristic length scale (l_{MAJ}). The l_{MAJ} may be interpreted, but is not directly associated with the SC coherence length (ξ).

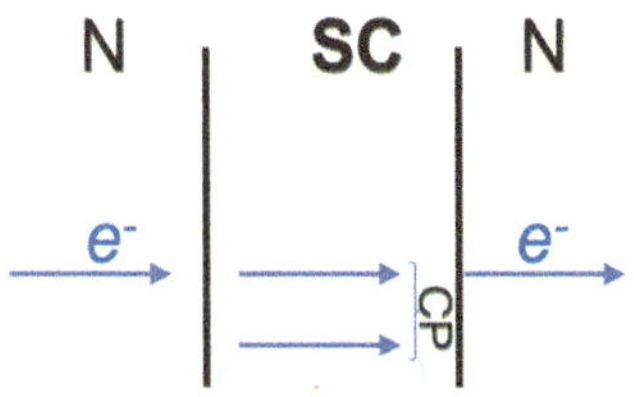

Fig. 2.2 *Andreev tunnelling process, yielding Andreev-bound states (ABS).* The incidence of an electron (e^-) from a normal material (**N**) onto a superconductor (**SC**) is accompanied by the creation of an Cooper pair (CP) and a charge compensating hole (h^+) in the SC. The possibility of a e^-–h^+-bound state (Andreev-bound state: ABS) is indicated through the oval. The ABS states may correspond to zero-energy states as well and may be conflated with the Majorana zero modes

2. Partially separated ABS (ps-ABS) separated by distances longer then l_{MAJ} but less than the length of the NW. These states may be considered as a superposition of two MBS separated by a distance smaller than their decay length.
3. A true MBS corresponding to the MBS localized at the ends, for example, of a NW.
4. The relative position of the states along a one-dimensional NW is illustrated in Fig. 2.3.

It was argued [53] that quantized conductance peaks which are robust to variations in system control parameters such as Zeeman field (E_z), tunnel barrier height, proximate potentials/voltage, etc. may not necessarily arise from topological Majorana modes alone, but may be related to the ps-ABS [52] with the same robustness to variations. In NW systems with lengths smaller compared to the decay lengths of the MBS located on its two ends, a proximity-based overlap of the wavefunctions of the putative MBS at the ends allows for interactions resulting in the splitting of the ground state and formation of the trivial ps-ABS. Local measurements unfortunately cannot be used to distinguish the two (MBS and ps-ABS) as they would have strikingly similar signatures, as measured in electrical transport. Indeed, only nonlocal measurements may clearly identify the topological nature of the states and hence distinguish ps-ABS and MBS.

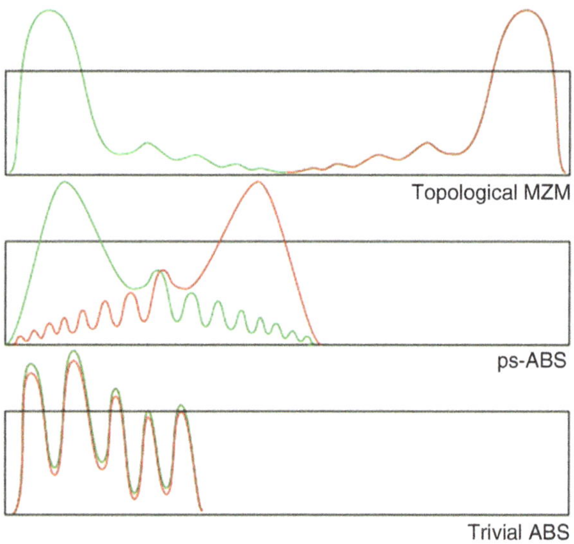

Fig. 2.3 A description of the wavefunctions related to the ABS and the Majorana zero modes (MZM) along a NW. The *left* (*/right*) Majorana mode is depicted as *green* and *red*, respectively. A partial (/substantial) overlap of the individual MZMs yield a ps-ABS (/trivial ABS). (From [193])

2.5 Mid-Gap States of the Andreev, Yu–Shiba–Rusinov, and the Coulomb Kind

The occurrence of defect states in the bandgap may considerably influence the related device characteristics, for example, in semiconductor-based systems [54, 55]. While the SC gap is generally robust to the influences of nonmagnetic impurities, in accordance with Anderson's theorem [56], there is yet an "extreme sensitivity" to magnetic impurities [57]. The latter variety along with the Andreev states and the Majorana modes, indicate the possibility of three types of sub-gap states. The study of such mid-gap states is of relevance as the relevant signatures may be confused with those of the sought after Majorana modes.

Generally, when a finite size SC is considered, for example, a SC island, there is also the possibility of charges being put on the island, say from a quantum dot (QD). There arises a possibility of adding electrons one at a time. While it is nominally expected that the electrons would be added or would participate in even numbers, condensing as Cooper pairs in the SC, there is a substantial possibility of a quasiparticle-flavored odd electron due to the E_c considerations. The spin from such an electron can induce a *localized excited state* in the SC energy gap, as was indicated by Shiba [58]. The interaction of such quasiparticle excitations, termed Bogoliubons [59], with electrons in the supplying QD would form singlet states and yield the sub-gap states termed the Yu–Shiba–Rusinov (YSR) states: Fig. 2.4b. In summary, we consider two other energy scales in addition to the Δ_{SC}, that is, the electrostatic charging energies of the SC island: $E_{c,SC} = e^2/2C$, with C as the island capacitance) as well as that of the QD: U. There then exist the following three regimes:

1. Andreev states: $\Delta_{SC} >> U$; $E_{C,SC} = 0$.
2. Yu–Shiba–Rusinov (YSR) states: $U >> \Delta_{SC}$; $E_{c,SC} = 0$.
3. Coulomb states: $E_{c,SC} \sim U, \Delta_{SC}$.

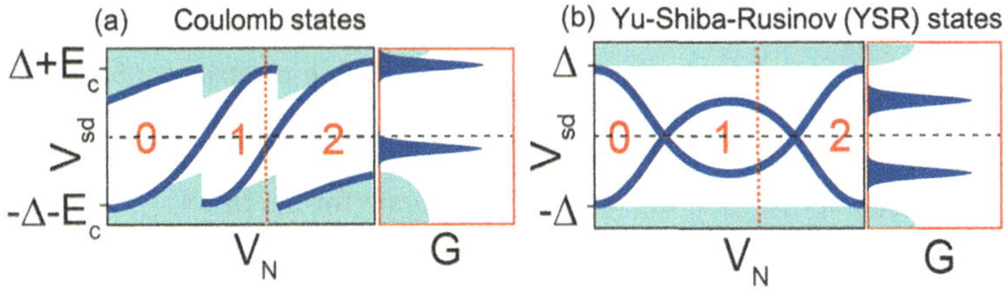

Fig. 2.4 The formation of sub-gap states in the energy gap of a SC (Δ), with added electrons from $0 \rightarrow 1 \rightarrow 2$. (**a**) Coulomb states are formed when the electrostatic energy, related to the SC island capacitance ($E_{c,SC}$) is of comparable magnitude to the Δ. There is a reduction of the electron–hole symmetry with the resultant formation of mid-gap states. The variation of the related *asymmetric* states with the voltage on the supply quantum dot (V_N) and the related differential conductance (G) is shown, with respect to the continuum background (*cyan*). (**b**) In contrast, when $E_c = 0$, the YSR states are formed symmetrically around the mid-gap, corresponding to the preservation of the electron–hole symmetry. There is then the possibility of a finite G even at the mid-gap

The Andreev states would be obtained in a limit, where there would no barrier to the addition of electrons to the SC island. YSR states would be obtained when there is such a barrier and only single-particle additions, for example, as a quasiparticle excitation, are feasible. However, when the $E_{c,SC}$ is comparable to or larger than the U in a SC, charge-based correlations between the SC quasiparticles and the QD electrons reduce the electron–hole symmetry characteristic of the SC and consequently yield mid-gap states. It is noted that the Andreev states could be substantially reduced when the finite size of the SC is considered.

Extending the notion of an isolated spin to a paramagnetic impurity would then imply the role of magnetic impurities in inducing such mid-gap/YSR states, as will be discussed further, for example, in Sects. 3.3.2 and 3.5.3. The rationale for their formation is related to an effective reduction of the SC gap from an *intrinsic* magnetic field related to the impurities. The in-gap states reflect the energy scales associated with the scattering of the Cooper pairs in a SC by the impurities. Generally, the G of the YSR states is much smaller compared to the ABS states considering their origin from scattering processes. There is also the possibility of Coulombic sub-gap states [60], when $\Delta_{SC} \sim E_c$ where the SC island interacts with the external electrons in the QD. The energy levels from the resulting charge distribution, with broken electron–hole symmetry, are indicated in Fig. 2.4a, with the two in gap energy levels at different energies relative to the mid-gap energy. It was noted that the Coulomb state–related levels could be closer to the mid-gap and be conflated with the Majorana modes through the coupling of a QD-like state to topological SC (with inverted energy band gap) [60].

Consequently, significant attention and discrimination is necessary for MBS identification, and the distinction is often unintentionally obscured, for example, through disorder in the material, contacting electrodes, etc. Other possibilities for the zero-energy states include contributions from geometric resonance effects, from oscillations in the density of states (DOS), of the Tomasch [61] type or the Rowell–McMillan [62] type. More generally, zero-energy states may always be present in trajectories where the SC order parameter is real [63].

We will now discuss key experiments which have claimed the identification of the MBS or Ising anyons [49] and their subsequent modulation. The initial focus will be on the experiments seeking to identify MBS in electrically contacted systems where carriers tunnel into the states. Subsequently, we will discuss alternative noncontact measurements, for example, those involving scanning tunnelling microscopy (STM) and optical probing, etc., to investigate for the MBS.

Chapter 3
Experimental Investigations of Majorana Modes and Majorana-Bound States (MBS)

It was seen in the earlier sections that *unique* zero-energy states suitable for quantum information science may be obtained in novel superconductor-based systems in topological materials/insulators. Much of the experimental work related to identifying TS has been performed in lower-dimensional systems and heterostructures [50, 64–67] such as one-dimensional (1D) nanowires, and involving both elemental and compound NWs. Typically, the latter variety with heavy elements, such as InSb, InAs, etc., have been preferred as they have a larger degree of spin–orbit (S–O) coupling. A typically reduced Schottky barrier for charge transfer with metal electrodes, through the use of Indium (In)-based materials, is also advantageous.

When 1D nanowires are coupled to SCs, proximity effects [41] induce SC states *in* the nanowire. Zero-energy states may be posited to exist at the point where the top and the bottom edges of the induced superconducting energy gap overlap as may be induced through an externally applied magnetic field (B_{ext}). Considering then both the (a) 1D characteristic, and (b) S–O-coupling-induced magnetic field polarity, a particle situated in a zero-energy state in the proximity-induced SC nanowire is simultaneously pushed to the two boundaries, and a measurement of the related superposition could be interpreted as a signature of the Majorana states/modes. However, such identification is quite error-prone, for example, zero-energy states may also be manifested from bound states arising via the constructive interference of the particle-like and hole-like quasiparticles, yielding Andreev-bound state (ABS) [68], and must be carefully carried out.

Consequently, as will be discussed in more detail below, there is *considerable ambiguity* as to whether the experimentally observed features are truly representations of unique TS amenable for storing quantum information and whether they can ultimately be used for QC [53, 69]. While specific SC types, for example, those exhibiting *p*-wave superconductivity (where electron spin is not important), such as Sr_2RuO_4, were initially [70] proposed for hosting the TS, later studies have focused on investigating the TS in topological insulators [41, 71] proximal to *s*-wave (with

© The Author(s), under exclusive license to Springer Nature Switzerland AG 2022

P. Bandaru, S. Natani, *Topological States for New Modes of Information Storage and Transfer*, https://doi.org/10.1007/978-3-030-93340-1_3

electron spin singlet) SCs. However, the proximity coupling generally requires long coherence lengths associated with low-temperature, small-gap SCs.

3.1 Majorana Modes in One-Dimensional Nanowires (NWs)

A one-dimensional semiconductor NW provides a natural testbed for probing the existence as well as the localization of topological states such as the MF-related modes. The 1D manifestation is a prototype of the system envisaged by Kitaev in his original proposal for the possible existence of the MFs [70]. A characteristic 1D system is subject to quantum confinement in the two dimensions transverse to the length. In present experimental practice, NWs are mostly synthesized through chemical vapor deposition (CVD)–based methods, for example, based on the vapor–liquid–solid [72] mechanism, where a crystalline NW nucleates and grows outwards from an liquid Au catalyst support subsequent to adsorption of the NW constituents from the gas phase. The "whisker" form of the NW abetted by slow growth ensures the minimization of defects, such as stacking faults. The NWs are then either collected from the substrate and sprinkled onto pre-patterned electrodes or contacts made to them in situ on the substrate. One-dimensional compound NWs and coupled to *s*-wave superconductors (SC), through proximity effects, enable SC concomitant states *in* the NW [73]. As indicated previously, zero-energy states corresponding to the MBS may exist at the point where the top and the bottom edges of the SC energy gap overlap. Such overlap is obtained through the following:

(a) Choice of material, which sets the *internal* magnetic field (B_{int}) from S–O coupling.
(b) Changes in the carrier concentration, through a tuning of the chemical potential/Fermi energy (E_F) as may be obtained through the use of external gates. Here, we invoke the accumulation (/depletion) of electrons (/holes) under positive (/negative) voltages. Such an aspect is the foundation of field effect transistors [54, 74] and has been extensively studied in the context of the semiconductor industry.
(c) Using an external magnetic field (B_{ext}) to match the internal energies arising from the B_{int} and the E_F.

The relative tuning of the three underlying parameters is then the basis for probing tunnelling from an electrical contact into the putative MBS.

From the 1-D aspect of the current flow, a conductance quantum of G_o is expected to be observed in the I–V characteristics [75], when current flows through a MBS. Generally, the experiments involve tunnelling of electrons across a barrier from a normal contact into the putative MBS in the SC system. It was previously discussed [76] that the particle–hole symmetry related to the MBS mediates a resonant Andreev reflection, with identical tunnel coupling for the electron and the hole in a given electrode facilitating the observation of a *perfect* G_o. Such a conductance peak for the zero-energy state is expected at zero-bias and termed a zero-bias peak

(ZBP). It will be seen that while the ZBP is necessary for the identification of a MBS, it is *far from sufficient.* Moreover, disorder and thermal effects on the ZBP intensity must also be considered, with respect to the values related to the expected conductance quantum [77].

3.1.1 Tunnel Barriers to Channel Electrons from a Contact into an MBS State in a NW

An instance of an experiment [50, 78, 79]aimed at finding the MBS involved transport of electrons across tunnel barriers whose barrier height was controlled through external gating. Further the chemical potential, that is, the E_F of the system was manipulated by either an intricate pattern of gates below the NW or through a common back voltage applied on the substrate: Fig. 1.3a. Typically, the NW is incompletely covered by the SC making precise control of the E_F, through gate voltages difficult. The key aspect to identifying a robust MBS would be the immunity of the identified conductance peak to external influences, over a range of magnetic fields as well as the gate voltages. The appearance and vanishing of the MBS-associated ZBP needs to be carefully correlated to the reversible phase change from a trivial SC state to a topological SC state.

3.1.2 Conductance Quantization to $2e^2/h$ ($= G_o$) in NWs as a Metric for an MBS

An important breakthrough in the fabrication of such NW–SC hybrids for probing MBS was related to the aspect of producing a *hard* SC gap [80] through eliminating known disorder related features such as disorder, roughness, etc. Such nominally inevitable features relate to a *soft* SC gap [81]with the prospect of a finite density of states (DOS) in the gap and may contribute to mid-gap conductance. The consequence is that the ZBPs, in the latter case, would be related to conductance values less than G_o. Better-quality materials with less disorder and better interfaces were sought for *hardening* the gap and hence reducing the quasiparticle density of states, in the gap of the topological SC. This breakthrough also made possible a better coverage of the semiconductor NWs with an SC, reducing the screening effects and allowing a more precise control of the chemical potential for the system.

In one instance [81], where such an aspect was claimed to have been accomplished, conductance quantization corresponding to transport through thin InSb nanowires provided evidence of current passage through a state ascribed to the MBS. The tunnelling experiments were conducted on a device constituted from the hybrid materials system (InSb NW of length ~ 1 μm covered partially by an Al superconductor): Fig. 3.1a. The partial coverage is to prevent screening of the

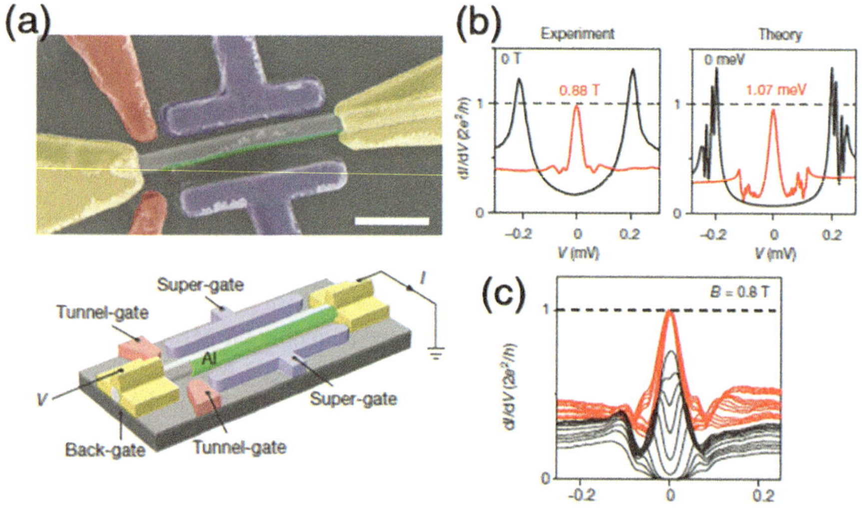

Fig. 3.1 *Tunnelling into a Majorana mode*. Adapted from [81]) (**a**) The device and materials system (InSb NW coated with Al) used for the demonstration of the quantized Majorana conductance of $2e^2/h$ (= G_o). The tunnel gates were used to set the barrier coupling between the *left* electrical contact and the SC Al while the super-gates set the electrochemical potential (μ), or the Fermi energy (E_F) in the segment covered by the Al. The *dI/dV* (differential conductance) characteristics were monitored and recorded (**b**) to yield a G_o at a B_{ext} greater than ~0.8 T. At such a field, the gap related to the proximity-induced superconductivity in the InSb NW is closed, yielding a zero-energy state related to the MBS. Lower magnetic fields induce coupling of the current to bulk modes in the NW. Reasonable agreement with theoretical predictions (*right*) was indicated, where the spikes in the *I–V* curve are indicative of the finite length of the NW. (**c**) The conductance quantization of the ZBP is robust over a set of tunnel gate voltages, but may yet be tuned corresponding to the interaction of the Majorana wavefunctions at either ends of the NW, and related to a peak splitting. For instance, increased tunnel gate voltages (shown) may be used to reduce the effective length of the topological SC–NW, and reduce the peak height

electrode-related electric fields, due to the SC. The experiments were conducted at 20 mK (yielding a thermal broadening, defined as 3.5 k_BT, of the order of ~6 μeV). Here, the tunnelling into the modes is sensed through *dI/dV* measurements and the tunnelling barrier height, proximate to the NW–SC interface, was set by the *tunnel gates*. The *super-gates* were used to change the overall chemical potential of the materials system. A ZBP (of width ~ 30 μeV, indicating tunnel coupling incorporating the thermal broadening as well as the lock-in bias excitation of ~5 μV) seems to be manifest at a B_{ext} greater than ~0.8 T: Fig. 3.1b. Alternately, at a B of 0 T, the *dI/dV* features correspond to current passage involving nonzero energy states. The experimental results were in reasonable accord with the theoretical predictions—on the *right* of Fig. 3.1b. The spikes in the theoretical simulations are because of the finite NW length. The nonzero differential conductance in the *hard* gap (with a voltage bias: $|V| < 0.2$ mV) was ascribed to robust Andreev tunnelling. As seen in Fig. 3.1c, the ZBP amplitude reaches G_o over several values of the tunnel gate voltages at a given value of the B_{ext} of ~0.8 T.

The ZBP quantization, typical of Majorana modes, was shown to be relatively robust to variation in tunnel coupling, magnetic field and chemical potential. The conductance of G_o persisted from ~0.8 T to ~0.9 T, beyond which a magnetic field–induced destruction of SC character was indicated. However, the ZBP was also claimed to be tuned through application of voltages from the tunnel gate or the super-gates—both of which influence the potential profile. The application of negative tunnel gate voltages was seen to enhance the interaction of the modes at the ends of the wire and after a certain limit cause a ZBP splitting: Fig. 3.1c. The splitting results in a depression of the ZBP while satellite peaks, on either side, are enhanced. A similar splitting and a reduction of the magnitude of the ZBP was also observed in the tuning of the chemical potential, through the *super-gates*. Tuning the voltages over a range of values can yield oscillatory behavior, corresponding to the overlap and subsequent repulsion of the energy levels, and causing a modulation in the magnitude of the ZBP, from G_o typically to smaller value. It was noted that a careful tuning of several gate voltages seemed to be necessary to obtain the G_o. The extreme sensitivity to the layout was seen through studies on a marginally smaller NW (of length ~ 0.9 μm), where conductance larger than G_o was seen and ascribed to transport through multiple channels from a more "open" tunnel barrier. Even here, while MBS was predicted to occur even in multi-sub-band wires, typically with an odd number of occupied sub-bands [82–84] and a sample width less than the SC coherence length, such aspects still remain to be understood in terms of discerning the individual contributions as well as their coupling to external electrodes. For instance, the closing of the SC gap to induce a topological state in the SC may be less well defined in this case. It was also noted that a temperature increase, from 20 mK upwards suppresses the ZBP conductance, which may involve contributions from such sub-bands.

While the correlation of such behavior to *quantized* and *split* ZBPs and the parameterization in terms of Majorana wavefunction overlap was interesting, such aspects demand a better understanding of the influence of voltage polarity on the nominally uncharged Majorana modes. Moreover, the related MBS interaction with the majority electron–hole population in the materials systems also needs clarification. The distinction between the MBS and the ABS, which may have energies close to/overlapping such modes, was discussed in terms of the propensity of the latter to *trivial* ZBPs with magnitude possibly larger than G_o, with respect to external tuning voltages (i.e., different behavior with respect to the back-gate voltage and the tunnel gate voltage) and magnetic fields. For instance, instead of binding to a zero-energy as would be expected in a MBS, there are level crossings observed for ABS-related states. It was indicated [85] that the consideration of realistic amounts of disorder in the InSb nanowire as well as the finite length of the NW, induces trivial Andreev-bound states with near zero energy which could also be robust to changes in chemical potential for a large range of magnetic fields. This is a veritable and nontrivial issue as increasing length and material volume would be expected to introduce more disorder, while possibly decreasing the multilevel character of the NW.

Moreover, these experiments seem to be considering tunnelling into only one Majorana mode/MBS at the end of the NW, while the other end is not apparently probed or subject to clear control. The experimental signatures are typical of a single-end state-coupling to the MBS. Additionally, while it is generally considered that Majorana modes exhibit less/zero sensitivity to external forces, in the absence of major perturbations, such aspects need to be better understood given that the modulation of electron population by the gates and related influences on the MBS is unclear in physical concept. Moreover, the parameter space for the robustness of the modes needs to be further clarified. Finally, even if the MBS and the related end modes were diagnosed and identified, the methodology related to the manipulation of such states for QIS is far from obvious.

In the previous experiment, tunnel gates were deployed to tune the tunnel barrier height, with the region between the tunnel gate and the super-gate corresponding to a reduced potential as would be the case in a quantum dot (QD). As the QD potential is modulated, the MBS-related ZBP is robust and bound to the zero-energy level while the ABS-induced ZBP would be indicated only at the potentials when the Andreev-energy-levels cross. The greater influence of the applied voltage on the ABS was posited to indicate a more closely coupled state in distinction to the MBS which corresponds to spatially delocalized modes. Here, tunnel coupling from the external contact is less influential compared to the internal ABS-state-related coupling. Consequently, there is the possibility that the ZBP width would be smaller and correlated to thermal broadening of ~6 μeV, compared to when MBS states dominate.

Considering that the MBS are formed in the limit of a low ABS density, provoked the exploration of tuning the number of ABS to yield Majorana mode–like characteristics. A hybrid QD–InAs NW system, where the coupling of the QD states with the sub-gap states in the nanowire could be investigated, was used [51]. An Al shell covering the InAs was used for inducing superconductivity into the latter: Fig. 3.2a, on the right-hand side of the NW, while Al is removed on the left-hand side of the NW. Consequently, a normal contact—QD–SC sandwich is formed along the length of the NW. The QD is hence a *spectrometer* of tunnelling related to the energy states, such as the ABS and presumed MBS in the proximal SC, with enhanced conductance observed whenever there is hybridization of the energy levels of the QD with the states in the NW. The connecting of the presumed MBS at the end of the wire with the QD can also be probed, for example, with respect to the leakage of the MBS wavefunction from the NW to the QD. Appropriate bottom gates (to control the electron density and chemical potential of the system) as well as top gates (to form a native QD in the nanowire as well as to tune the electron current passing between the QD and the nanowire) were deployed. The electron concentration of the QD and the InAs were controlled through separate gates to prevent mixing, induced through cross-capacitive coupling.

Monitoring the differential conductance (*dI/dV*) through the system, as a function of applied *B*, several sub-gap states were observed which were displaced to lower energies—incorporating a ZBP, at sufficiently large *B*: Fig. 3.2b. The particle–hole symmetry could be ensured through the observation of a symmetric spectrum around zero V_{SD}. There was seen, for example, through a decreased V_{SD} regime with zero conductance, a *softening* or reduction of the Al-induced SC gap from

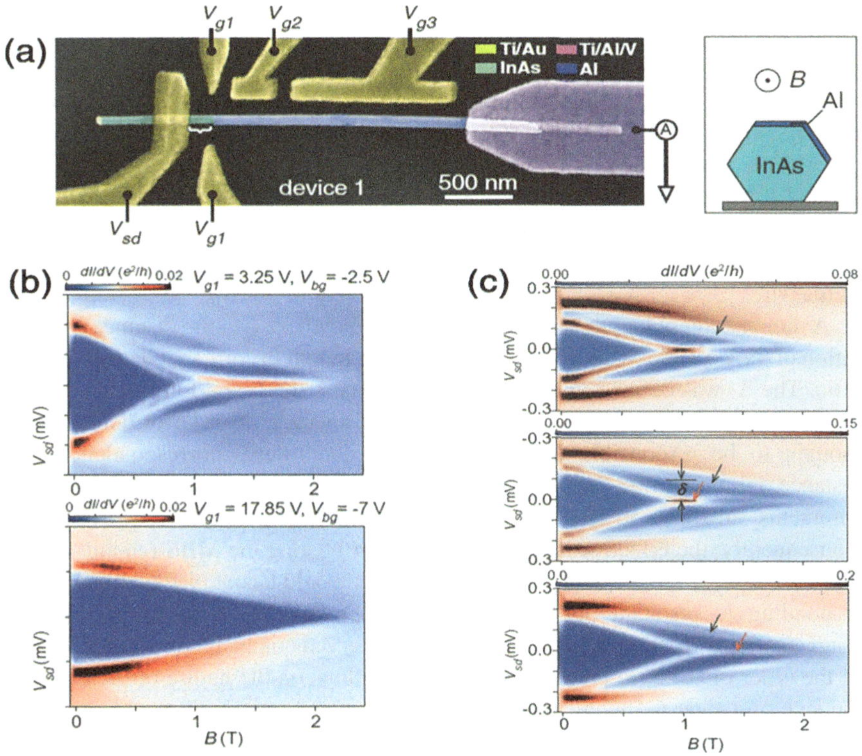

Fig. 3.2 *Identifying Majorana-bound states (MBS).* (Adapted from [51]): (**a**) Electrode scheme for identifying Majorana-bound states (MBS) in a quantum dot (QD) constituted nanowire (NW). *Left*: a false color micrograph of the scheme where the green wire represents the NW, and the blue part represents the overlaid Al shell. There is a confined QD denoted by the white bracket between the normal and the SC parts of the NW. The various gate voltages (denoted by V_g) are used for tuning the electron density as well as the chemical potentials. *Right:* The cross-section of the epitaxially grown InAs nanowire covered partially with Al. (**b**) The variation of the differential conductance (*dI/dV*), through the NW, as a function of the applied magnetic field ($B_{ext} = B$) and voltage across the nanowire (V_{sd}), measured at different top-gate (V_{g1}) and back-gate (V_{bg}) values. There is a larger density of sub-gap states, in the *top* figure, compared to the *bottom* figure, and scanning through these states yields enhancement in the conductance. The sub-gap states move lower in energy as the B_{ext} increases. (**c**) The differential conductance as a function of the B and V_{sd} in the low ABS density regime, as a function of increasing V_{g2} and V_{g3} (from top to bottom). The *top figure* corresponds to a splitting and re-merging of the ABS states, the *middle figure* shows the formation of a residual gap (δ) in the ABS due to the finite length of the wire around the trivial *to* nontrivial topological transition, while the *bottom figure* indicates a non-splitting behavior characteristic of the formation of an MBS. The ZBP corresponding to the MBS is indicated by the red arrow while the first excited peak corresponding to an ABS is indicated by the black arrow

~0.22 meV to ~0.20 meV, as a function of the increased ABS density. Moreover, the ABS energy would vary and could be tuned through the gate voltages and also be a function of the applied B. Tunnelling spectroscopy from the QD into the ABS were used to probe the sequential splitting, movement, and merger of the ABS states as a function of the *B*, which was ascribed to trivial parity change, However, an immunity to the external *B*, as manifested through a *sticking* at zero energy (over a range of *B*), was used for marking the evolution toward an MBS: Fig. 3.2c. The sticking effect was interpreted as due to the spin–orbit interaction induced anti-crossing of the ABS which results in a pushing/rearrangement of the energy levels at zero/mid-gap and the subsequent immunity to external environmental influences, such as the applied *B*. While the influence of the SOI is considered through an effective energy gap, the separation of the first discrete excited state from the zero-energy level is another distinct value ($=\delta$) of the order of 0.1 meV and arises from the finite length of the wire.

A dimensionless length ratio parameter ($= L/\xi$), where L is the distance of separation of the possible MBS and ξ is the coherence length in correspondence to a Δ'/δ ratio. The Δ' was defined to be an *effective gap* near the trivial to nontrivial phase transition and was estimated to be ~0.18 meV. However, disorder or defects were thought to be responsible in obtaining good quantitative correspondence. For instance, taking ξ to be ~260 nm and the energy gaps from the experimental measurements, an L of <50% of the nominal/overall length of the NW was obtained. Consequently, the effective length of the NW for hosting the MBS remains uncertain. Moreover, in the regime when the MBS state could be coupled to the QD, corresponding to resonant dot–wire coupling, the ZBP was immune to external magnetic fields from ~0.75 T to ~2 T, but the observed conductance of the ZBP was of the order of 0.1 e^2/h, precluding clear conclusions on the nature of the state, cf. the lack of a clear G_o quantization, which as indicated previously could be a measure of tunnelling into the MBS. There still remains then an issue of whether the clear resolution of the ABS, with respect to MBS formation, is feasible.

There was also severe criticism [69] that sticking of the ABS, with increasing B_{ext} does not itself imply the formation of the MBS, and may also be present whenever there is S–O coupling and Zeeman fields. Indeed an earlier study [50]: Fig. 3.1, marking the "signatures of Majorana Fermions" adopted a similar notion where the observed bound states were found *sticking* to zero energy over a range of B_{ext} of ~0.3 T, and yielding a ZBP conductance of ~0.05 G_o. With further increase, subsidiary peaks located somewhat symmetrically on either side of the ZBP and classified as ABS were seen: Fig. 3.2b. Such sticking, even in the presence of external gate voltages, for example, through varying the voltage bias over specific gates—see Fig. 3.3a, by a few volts was observed. As indicated by the data in Fig. 3.2c, obtained by varying the B_{ext}, the aspect that the ZBP is only manifested at a finite magnetic field and *sticks* to the zero-bias position is apparent. However, the magnitude of the ZBP value is of the order of 0.3 G_o or smaller. As indicated previously, the parallel contributions from multiple sub-bands may be involved, especially when the carrier concentration is of the order of 10^6/cm [79], which complicates the interpretation. There was also an intriguing suggestion that the coupling between the Majorana modes could reduce the observed conductance below G_o as well [79].

Indeed, such lower G_o were also observed in other NW-related studies as well, as indicated further in Fig. 3.3d [79] and Fig. 3.3f [78], related to the use of InAs and InSb-based NWs, respectively. In the former case, the ZBP is only stable in the

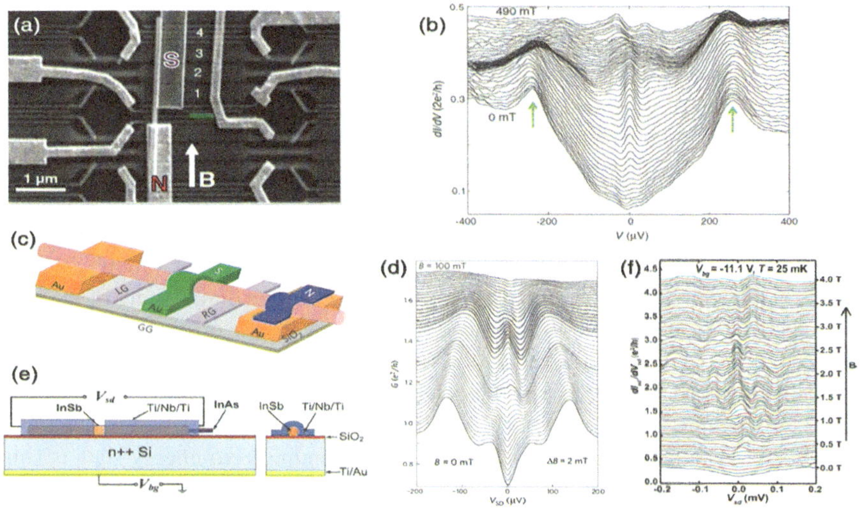

Fig. 3.3 *Zero-bias peaks related to the MBS* (**a** & **b** adapted from [50]) (**a**) Scanning electron microscope image of the implemented setup with normal (N) and SC contacts on top of an InSb NW. The SC contact partially covers the right part of the NW. The gates numbered 1,2,3,4 are used for controlling the chemical potential of the NW. The green rectangle schematically indicates the tunnel barrier related to the separating of the normal contact and the SC contact. (**b**) Differential conductance (*dI/dV*) versus bias voltage ($V_{SD} = V$) maps at increasing magnetic fields (0–490 mT in 10 mT steps). Each curve was offset for clarity. Zero-bias peaks (ZBP) of height ~ $0.05G_o$ were observed between B ~ 100 mT to ~400 mT. The induced SC gap of $2\Delta_{SC}$ is marked by the green arrows at the left and the right. (**c** & **d** adapted from [79]) (**c**) The setup for the device, an Al-InAs nanowire suspended on two gold contacts above a *p*-type silicon layer (which serves as a global gate (GG)) coated with 150 nm SiO_2. An Al SC contact at the center and two narrow local gates (RG and LG), 50 nm wide and 25 nm high, displaced from the SC contact by 80 nm, influence both the tunnel barrier height near the Al edge and the chemical potential in the wire. (**d**) A conductance vs. applied voltage (applied to SC contact) map at increasing magnetic fields (0–100mT in steps of 2mT). An inner induced SC gap at ~45 µeV and an outer SC gap at ~150 µeV, corresponding to the SC gap of bulk Al is observed. Each curve is offset for clarity. A ZBP can be seen for B_{ext} in the range of 35mT to ~70mT, beyond which the ZBP splits. (**e** & **f** adapted from [79]) (**e**) The electrode experimental setup (side view: left panel, cross-section view: right panel). The InSb segment of a InSb/InAs heterostructure with two Ti/Nb/Ti (3 nm/80 nm/5 nm) contacts at each end placed on a SiO_2-capped highly doped Si layer. A bottom layer of Ti/Au serves as a back gate (BG). (**f**) A *dI/dV* variation with applied voltage (V_{SD}) as a function of the B_{ext} (in the range of 0–4 T). The curves are offset for clarity. A ZBP can be seen to be present in a B_{ext} range of 1–2.5 T. The evolution of the conductance seems to indicate multiple opening and closing of several gaps, perhaps indicating the sweep through sub-gap energy states

range of 30 mT–70 mT, beyond which a splitting is observed. However, the ZBP was associated with B_{ext} orthogonal to the internal B_{SO}, through a diminished peak magnitude with increasing angle between the NW and the B_{ext}. The limited range of the B_{ext} over which the ZBP appears prompted the consideration of alternate mechanisms for the observed phenomena, such as the presence of sub-gap states alternate to the MBS. A small *g*-value may explain the limited range of the B_{ext} over which the ZBP is stable. Alternately, with the use of InAs NWs, the ZBP persists over a B_{ext} range of 1.5 T—as indicated in Fig. 3.3f [78]. While there is a correlation of the E_z ($=g\mu_B B_{ext}$) with respect to the SC gap (Δ_{SC}), as in Fig. 3.2b, through which a sweep of the B_{ext} is used to access and stabilize a unique zero-energy state, the limited range of the B_{ext} over which the ZBP is stable seems to indicate a limited stability of the state. Such lack of robustness of the state may be related to the observation that the observed ZBP is not directly linked to the Majorana mode. While a gradual quenching of the SC state, leading to a reduced supercurrent and lower conductance, itself may be more of an issue with higher B_{ext} as in [78], the ~40 mT stability range in [79] may be due to the sub-gap states. Further, the diminishing of the ZBP at higher B_{ext} would bring forth alternate mechanisms for the observed phenomena, for example, Kondo-type correlations, related to an increased resistance from exchange scattering between conduction electrons and magnetic impurity atoms [5].

The robustness of the ZBP itself is dependent on the magnitude of the bandgap in which it is situated, that is, whether the corresponding zero-energy state is situated in the gap corresponding to Δ_{SC} or at E_z. A larger (/smaller) gap leads to stable (/splitting) character of the ZBP. As indicated in Fig. 2.1d, typically a large E_z could be obtained implying that a more stable ZBP could be obtained through its tuning through the μ (with a gap at the center of the BZ).

As indicated previously, it is plausible that MBS may not be forming at the ends of the wire, say due to pinning by defects, etc., which would then reduce the magnitude of the ZBP. The interaction between proximal MBS would then lead to an energy splitting. Indeed, an exponential decrease in the energy level splitting, where an order of magnitude reduction for a micron of increased NW length was indicated [86]. Here, Coulomb blockade (CB) spectroscopy involving InAs NW segments in a range of lengths (0.3 μm to 1.5 μm) proximally coupled to SC Al was deployed. In the CB regime, corresponding to single electron transport from a contact into the NW island, and then to the contact, the small size and capacitance (C) of the island implies a large charging energy ($\sim e^2/C$) and clear distinction of the constituent energy levels. Consequently, the number of electrons on the NW would be regulated by its size. At the smaller length scale, there is a larger energy separation compared to that at larger length scales. If a NW is constituted from two MBS at the ends, the transport would proceed through the hybridized energy levels. A larger B_{ext} would also modulate the transport to occur through the lower energy level. Such energy-level spectroscopy to distinguish the energy levels and their separation was performed. It was seen that there was an exponential decrease in the energy level spacing with increased NW length.

The fabrication of the related materials for probing such CB modulated transport is based on epitaxial growth of a thin Al layer (~10 nm) on two facets of the hexagonal section of the InAs: Fig. 3.4a. Suitable electrode contacts were made to the NW-constituted islands and CB transport probed. The gate voltage (V_G), beneath the NW, is used for tuning the energy, and related transport of the electron from the left contact into the NW and then subsequently into the right contact. The experimental observations encompassed the periodicity of the conductance and is related to whether the transport involves two *or* one electrons, the latter arising from the

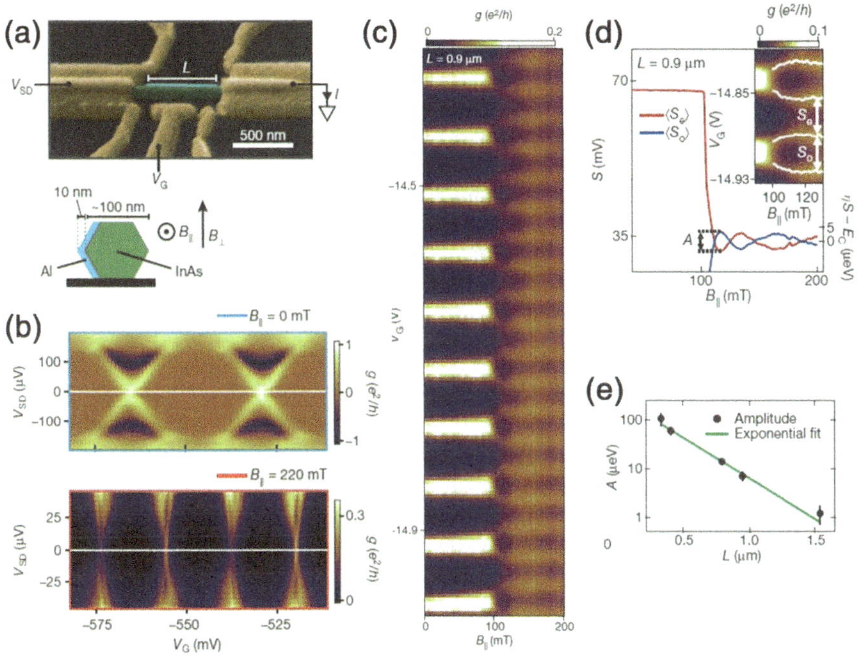

Fig. 3.4 *The probing of the interaction and separation of the MBS in a NW.* (Adapted from [86]) (**a**) (top) Image (in false color) of the electrode setup. Two yellow Ti/Au contacts on each end; a green epitaxial InAs nanowire; a light-blue epitaxial two-facet Al shell (of length L). V_{SD} is the applied voltage bias across the device and V_G is the gate voltage to control the chemical potential of the nanowire. (bottom) Cross-section of the InAs nanowire showing the orientation of the Al shell and field directions B|| and B⊥. (**b**) A map of the differential conductance as a function of the V_{SD} and V_G maps at two different parallel magnetic fields (0 mT and 220 mT) indicating a series of diamond-shaped modulations characteristic of Coulomb blockade (CB). A doubling of frequency of the diamonds (*corresponding to topological phase transition*) can be seen in the two distinct cases. (**c**) Zero-bias conductance plotted as a function of B_{ext}, applied parallel to the NW and V_G, indicating the transition from two-electron-related Coulomb peaks (related to the *trivial* SC phase) below 100 mT to nearly one-electron Coulomb peaks (related to the *nontrivial* SC phase)) above 100 mT. (**d**) Average peak spacing for even and odd Coulomb valleys $<S_{e,o}>$ from the measurement shown in (**c**) as a function of parallel magnetic field. The Coulomb peaks spacing oscillates around $<S_e> = <S_o>$ after a B_{ext} of ~110mT. Inset, high-resolution measurement for $L = 0.9$ μm. (**e**) The plot of amplitude of oscillation: A—in part (**d**) of the $<S_e>$ for five devices with different nanowire lengths. The green line is a fit to the data $A = A_0\exp(-\xi/L)$ with ξ as the SC coherence length, indicating an exponential increase in the robustness of the MBS with increasing NW length

transitioning of the SC system from a trivial to a topological phase: Fig. 3.4b. The *diamond-like* shapes, in the energy–charge space (V_{SD}–V_G) here, are related to negligible current inside the diamonds and finite conductance along the four lines (corresponding to the four modes of transport into/out from the left/right contact) of the diamond [87]. The periodicity of the zero-bias conductance as a function of the V_g and the B_{ext} is illustrated for the $2e^-$ periodic Coulomb peaks in Fig. 3.4c. Further, the voltage differences between the individual conductance in the topological phase oscillates with increasing magnetic field: Fig. 3.4d. Such oscillation was related to the hybridization of the energy levels of the MBS pair. It was indicated that while even peak spacing $<S_e>$ is obtained at $B_{ext} < 100$ mT, at a larger B_{ext}, there is an evolution into the odd peaks (corresponding to the transition to the topological phase), and an oscillation with increasing periodicity. Moreover, the amplitude of the oscillation (A) in Fig. 3.4e, is seen to decrease exponentially with increasing NW length (L), as $e^{-L/\xi}$, with an interaction length scale set by the SC coherence length: ξ—estimated to be ~260 nm. The modulation of the interaction, as related to the overlap of the MBS at the ends, is evidenced in a NW with $L >> \xi$. The exponential reduction in the A may be related to the corresponding *protection* of the MBS states in a longer NW and agree with the expectation that the MBS should be infinitely far apart for the emergence of a true, non-interacting Majorana mode.

However, it was argued [88] that there seems to be an anomaly in the observed experimental results in contrast with nominal understanding. One such example was the magnetic field dependence of the oscillation amplitude. It is well known that for a SC an increase in the B_{ext} would result in an increase in the ξ as well. This would imply that with a constant NW length, increasing the B_{ext} would result in the *increase* of the amplitude of the oscillation, while the experimental results in Fig. 3.4e seem to indicate the opposite, that is, a decrease. It was further hypothesized that such a decrease in amplitude would take place at high temperatures and in a system consisting of *both* ABS and MBS. For instance, if the SC covered the central part of the NW, then the ends would still permit ABS, the overlap of which was posited to explain the experimental results. Moreover, there could have been unintentional QDs through the NW length that could have contributed to the CB-related observations. It was further argued that the experiment simply reflected the physics of ABS in finite-sized NWs and could not be construed in terms of a clear signatures of MBS. While higher magnetic fields could be more reflective of MBS, the use of Al implies a reduction of the SC gap quite close to the point at which the MBS would be operative.

In short summary of the possible identification of MBS through NW-based transport measurements, no clear signatures have been obtained. The invasive nature of the investigations, that is, both through transport through the MBS as well as external magnetic fields, combined with a clear lack of the distinction between low-energy ABS and higher-energy MBS are contributing factors. An alternate way of probing the MBS harnesses the Josephson effect, which may be manifested even in the absence of an external bias field, for example, through the development of a voltage arising from a phase difference in the wavefunction [8].

3.2 Identifying Majorana States Through the Josephson Effects

The manifestation of the MBS through their influence on the phase is at the origin of probing through the Josephson effect. Here, the tunnelling of Cooper pairs from one SC, through a barrier, into another SC forms the basis for the Josephson effect

[21]. The related Josephson junction (JJ) may be constituted from a SC placed in between two non-SC (normal metals or insulators) contacts or vice versa. In one manifestation, just the phase difference (δ) between the right and the left junction, for example, between the SC and the non-SC, drives a constant current: $I = I_o\ Sin\ (\delta)$, with I_o related to the pair-coupling and the transfer of the Cooper pairs across the SC. Such a current flows through the JJ even in the absence of a biasing voltage and the related effect has been termed the "DC Josephson effect." However, we have previously seen that in conditions involving the MBS, there would only be a single electron mid-SC gap state. Consequently, tunnelling through a topological SC where such a state is possible involves a single electron. The related phase difference across the junction would be reduced by a factor of two. An alternate equation: $I = I_o\ Sin\ (\delta/2)$, yields a 4π-related phase difference (following the periodicity of the sine function), and has been termed the *fractional* Josephson effect [89]. The period doubling from 2π to 4π could, in principle, be discerned in the electrical current signal, and yield a signature for the MBS-related single electron state formation. The periodicity of the electrical current oscillations is in correspondence to the related physical processes, that is, the 2π related to the Cooper pairs (which carry the bulk supercurrent) while the 4π would be related to the Majorana modes. Indeed, the suggestion for the use of Josephson currents for diagnosing Majorana states at the ends of the NW arose from the original proposal of Kitaev [34].

Alternately, the AC Josephson effect considers a frequency-dependent modulation of the current through the junction, in the presence of an applied constant bias (V_{bias}) across the JJ. Then, the $I_{AC} = I_o\ Sin\ (\delta - \phi)$, where ϕ ($= \frac{2eV_{bias}}{\hbar} t$) represents the additional phase accumulated over a time (t). A related phenomenon, which forms the basis of the voltage standard in use today, is then the *inverse* AC Josephson effect which describes the behavior of the junction subject to electromagnetic (EM) waves. Given the coherent state of the Cooper pairs (with an effective charge = $2e$) in a SC, phase locking with the EM waves would be expected and recorded at specific voltages (i.e., V_n) in the I–V characteristics of the JJ when $(2e)V_n = n\ hf$. Here, n is an integer, and f is the frequency related to the incident EM wave frequency. Such constant voltage Shapiro steps would be modulated in the presence of Majorana modes which are now related to an effective charge of e and *not* $2e$. The concomitant frequency would then be halved to eV_n/h. This would imply that all the odd Shapiro steps would be absent in JJ based I–V measurements, involving single-electron MBS states.

From an experimental point of view, when a topological SC wire is placed in between two non-SC contacts, the current in the topological SC would be carried by both single-electron tunnelling and two-electron Cooper pairs. While the latter aspect is well understood in SCs, the former characteristic would arise from Majorana state–involved single-electron levels. Andreev reflection and tunnelling involving single as well as Cooper pair electrons, as indicated in Fig. 2.2, would be involved and yield sub-SC gap ABS as well. The possible MBS would then be comprised of sub-gap ABS, and would, in the infinite barrier JJ case, degenerate to *zero-energy* mid-gap states. Such bound states in the topological SC may be numerically modeled through the Bogoliubov-de Gennes (BdG) equations [20] with appropriate boundary conditions [90]. Experimentally, the evidence of such ABS was indicated through a finite "excess current"—manifested at zero-voltage biases [90, 91]. Fundamentally, the transfer of one electron, and the related measured current, *through* such zero-energy states would be proportional to the electron tunnelling amplitude. In contrast, the related current without such zero-states would be proportional to the *square* of the electron tunnelling amplitude in a *s*-wave SC. Moreover, the tunnelling time into and out of the zero-energy state should in principle be smaller than inelastic relaxation times relevant to the materials.

An alternative viewpoint considers that the *net* MBS mode corresponds to a superposition of the in-phase and out-of-phase component of a single-electron state (in the SC), with the related states placed at the end of the wire, as indicated in Figs. 1.2 and 1.3. Considering electromagnetic duality principles, each of these components should be associated with a magnetic field. To this end, we consider the Maxwell relation, related to the divergence of the magnetic flux density (***B***), based on the consideration that magnetic effects only occur only in pairs (as in a bar magnet or for the two MBS):

$$\nabla . \boldsymbol{B} = \boldsymbol{0} \tag{3.1}$$

From elementary vector analysis, we have, $\boldsymbol{B} = \nabla \times \boldsymbol{A}$, with $\boldsymbol{A}$ as the related vector potential. Taking the area (**S**) integral of both sides, and through the Stokes theorem—from $\oiint \nabla \mathrm{X} \boldsymbol{A} . \boldsymbol{n} . d\mathbf{S} = \oint \boldsymbol{A} . \boldsymbol{d}\mathbf{l}$, we obtain for the magnetic flux (Φ), using the relation $(\Phi) = \boldsymbol{B} \cdot \mathbf{S}$:

$$\Phi = \int \mathbf{A} . \mathbf{d} \, \mathbf{l} \tag{3.2}$$

As is well known, the phase (ϕ)–wavevector ($\boldsymbol{k}$) relation may also be depicted through a line integral of the form:

$$\phi = \int \mathbf{k} . \mathbf{d} \, \mathbf{l} \tag{3.3}$$

Equating the field momentum ($e\boldsymbol{A}$) to the kinetic momentum ($\hbar \boldsymbol{k}$) [21, 92] we obtain, from the ratio: $\boldsymbol{k}/\boldsymbol{A} = e/\hbar$, and the flux–phase relation, in units of

$$\Phi = \frac{h}{2\pi e} \phi = \Phi_0 \frac{\phi}{2\pi} \tag{3.4}$$

Here, the flux may be considered quantized in terms of multiple (h/e) units—termed a flux quantum ($= \Phi_0$), of ~$4.1 \cdot 10^{-15}$ T.m^2. Such a link is particularly relevant in materials, for example, such as superconductors or topological systems, subject to quantization rules where ϕ may only change in discrete values implying flux changes in discrete values. It is then apparent that with $\phi = 2\pi$, the related flux would be Φ_0—implying the association of a flux tube with the electron. However, in the case of an MBS, comprised of *half* of an electron state, we need a phase difference $\phi = 4\pi$, that is,

$$\Phi = \frac{h}{2\pi e}4\pi = \frac{h}{\left(\frac{e}{2}\right)} \tag{3.5}$$

Consequently, every 4π change of phase would be equivalent to adding one flux quantum corresponding/binding to one electron in the topological SC. The phase difference must be a minimum of 4π, to add two MBS corresponding to one electron state in a topological SC. We will next see how the phase degree of freedom may be deployed to possibly identify the MBS. Most experimental arrangements deploy an interferometric configuration involving JJs embedded in two branches of a loop, that is, a Superconducting Quantum Interference device (SQUID). The modulation of the phase in a *test* device in one branch of the loop by external forces, that is, radiation, electrical or magnetic field bias, with respect to a *reference* JJ in the other branch may then be detected with extreme sensitivity. Indeed, magnetic field measurements of the order of femtotesla (~10^{-15} T) corresponding to exquisite phase sensitivity seems routine [32].

3.2.1 *The Phase Variation for Deducing MBS in a Topological Superconductor*

From the phase (ϕ)–flux (Φ) relation for a SC: $\phi = 2\pi\frac{\Phi}{\Phi_o}$—see Eq. (3.4), with Φ_o (= $h/2e$)—with $2e$ as reckoned for Cooper pairs, it may be inferred that biasing the junction with a phase of π, corresponds to the addition of a flux of $\frac{\Phi_o}{2}$. In a trivial SC, this is not possible while in a topological SC, this phase bias corresponds to the addition of a single electron, with the related formation of two Majorana modes/MBS. Such a phase bias may be tuned through a magnetic field perpendicular to the Josephson junction ($B_\perp$), which introduces the flux into the loop. It is then expected that in a topological SC (induced by B_{ext} as discussed earlier—Sect. 2.1), that a ZBP would be manifested as a function of the $B_\perp$corresponding to when $\frac{\Phi_o}{2}$ is introduced, and related to the conduction current mediated by the single electron. The use of $B_\perp$also yield another tuning knob, additional to the B_{ext}, for modulating the characteristics of topological SCs. One example of such an experiment related to inducing SC into an InAs-based two-dimensional electron gas (2-DEG), through proximal Al gates[94]. A SQUID loop arrangement (through which the $B_\perp$was threaded) was deployed with two distinct JJs in the two arms—Fig. 3.5a. When a B_{ext} is applied parallel to the JJ surface (and perpendicular to the intrinsic B_{SO}), the InAs is tuned into a topological SC phase, with a single-electron state, and may induce two MBS at the ends. The tunnelling of electrons into the top JJ to probe a Majorana mode was regulated through a quantum point contact (QPC)—which defines electron puddles and appropriate gate voltages. The junction region, where the Majorana modes could be observed was patterned in a one-dimensional line, the width (~100 nm) of which

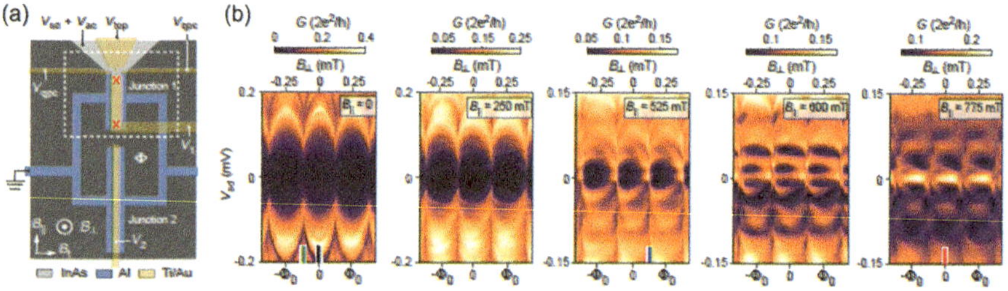

Fig. 3.5 *Phase tuning in a Josephson junction for MBS.* (Adapted from [93]) (**a**) A SQUID based on the arrangement of two JJs (Junction 1 and Junction 2). The phase difference across Junction 1 can be tuned through an applied flux (Φ, corresponding to a magnetic field perpendicular to the SQUID loop: $B_{\perp}$). The $B_{//}$ corresponds to the external magnetic field (B_{ext}) which tunes the Zeeman gap (E_z).The Ti/Au top gates V_1 and V_2 (isolated from the Al by HfO_2 dielectric) are used to control the chemical potentials/E_F of the junctions. Proximal superconductivity is induced from the Al top gates into the InAs. A two-dimensional electron gas (2-DEG) is shown in the Junction 1 region, at the ends of a line of which are the possible Majorana modes (represented as two X). The quantum point contacts (QPC) at the top is used for confining the 2-DEG, and the tunnelling into the modes is configured through the top gate contacts (V_{top}). (**b**) The conductance variation along the channel, as a function of $B_{\perp}$(which tunes the Φ) with increasing $B_{//}$ ($/B_{ext}$) – from 0 mT on the *left* to 775 mT on the *right*. The periodic oscillation of the SC gap with $B_{\perp}$along with the substantial enhancement of the conductance with $B_{//}$ is observed. The aspect of a wide topological phase region, on the right, at a $V_{sd} = 0$, when the $B_{\perp}$is tuned to a phase of π corresponding to $\frac{\Phi_o}{2}$ is remarkable

was much smaller compared to the ξ. The electrode scheme also allows for the tuning of the carrier concentration/E_F. The electron temperature in the devices was of the order of 40 mK. It was observed that as the B_{ext} was increased, the differential conductance: G (related to the magnitude of the current as a function of the V_{sd}, and recorded through standard lock-in technique, with an AC bias superposed on top of a DC bias, $V(t) = V_{sd} + V_{ac} \sin(\omega t)$, with V_{sd} ($/V_{ac}$) the order of 0.2 mV (/3 μV)) was increasingly dense close to $V_{sd} = 0$, and ascribed to the population of the SC gap by ABS. Beyond a critical B_{ext} of ~0.75 T, the ZBP that may be related to the Majorana mode is evident—see the lighter color at $V_{sd} = 0$, in the last panel of Fig. 3.5b. More quantitatively, the related ZBP was characterized through a negative curvature of the G (written as G'') at $V_{sd} = 0$, which was monitored through recording the third-harmonic component of the I_{sd}. The underlying notion is that a positive (/negative) G'' implies a trivial (/nontrivial) SC phase, where the G nominally increases (/decreases) with bias. The range of the ZBP stability, with respect to the B_{ext} was subject to an approximate lower limit of ~0.75 T and upper limit of 1 T, the latter due to the expected diminution of the SC gap. The sensitivity of the ZBP to tuning of the chemical potential (through external Ti/Au metal gates separated from the InAs by ~15 nm of HfO_2 dielectric) was probed over a range of ~4 mV, as well as through a 70 mV tuning of the gates connected to the QPC.

In this experiment, careful tuning of the E_z (at ~0.1 meV) as well as the μ (at ~79 meV) seems to be necessary, to harness the sensitivity of the topological SC to phase tuning. Additionally, the finite size of the system (width as well as the length ~ 1.6 μm) as well as the related fragility of the 2-DEG, such as the finite thickness, may also play a role. For instance, a large E_z was expected to reduce the phase-induced gap as would be manifested through the hybridization of the putative MBS states at the ends (marked by the X in Fig. 3.5a over the finite length. The rationale for positing the negative value of the G'' as evidence of the Majorana-based ZBP was not clear, in addition to a firm understanding of the observed values of the G (@ $V_{sd} = 0$).

The experiment involves yet another method of interpreting the topological phase and a related transition from the nominal phase regulated by the E_z alone. Here, nominal diffusive transport is associated with single particle motion, and the transition from a E_z-dominated regime to a diffusion-dominated regime may be related to a topological phase transition. Subsequently, yet another energy scale related to the Thouless energy (E_T), relevant to the flux-dependent single electron current [94], must also be considered. Generally, the E_T is associated with diffusive conductors, that is, corresponding to particle-like transport, with an effectively infinite wavelength and corresponding zero-energy. However, when the E_z is comparable to E_T, a topological transition related to the JJ is expected. A change in the phase: $\Delta\phi$ (= $2\neq\frac{E_z}{E_T}$), due to the $B_\perp$was defined [95], where the numerator (/denominator) may be related to the external (/coupling) flux. In this case, a phase change of π (/2π) would be related to $E_z = ½\, E_T$ (/$=E_T$). In the earlier study [95], the E_T was estimated at ~2.8 meV, an order of magnitude larger compared to E_z, which implies that there was no particular transition to the zero-energy/diffusive regime. However, the possibility of inducing the E_z to be comparable to the E_T, through changing the ϕ by $B_\perp$ as well as B_{ext} was explored in a setup similar to the one discussed earlier, that is, in a 8 nm wide HgTe quantum well-related 2-DEG (carrier density ~ $10^{11}/cm^2$ and mobility ~400,000 $cm^2/V.s$), with induced superconductivity from Al leads deposited on top [95]—all arranged in a JJ configuration. With reference to in Fig. 3.6b, the $B_\perp$ is applied through the flux loop (as indicated by the upward pointing arrows in the z direction) and tunes the ϕ. Along the proximal SC channel in the x direction, the k_x is translationally invariant and the supercurrent was considered to be carried by bands of ABS, formed by Andreev reflections at the normal–SC interface in the HgTe. When the parallel field—B_{ext} is zero, then the ABS cross at $\phi = \pi$ at $k_x = 0$ (corresponding to the SC mid-gap) as seen by the crossing of the *dotted lines* in Fig. 3.6a—which may be interpreted as the electron–hole superposition in the normal/non-SC region or as the introduction of $\frac{\Phi_o}{2}$. However, when B_{ext} is finite, the spin states are resolved with respect to the phase, that is, the B_{ext} is in-phase (/out-of-phase) with respect to the spin-directions and the respective ϕ are shifted positively (/negatively) at the mid-gap. The range of ϕ values corresponding to the difference of the phase shifts were interpreted as relating to the formation of the topological SC phase. Here, at zero-energy, there is no intrinsic spin symmetry and the relation to ABS implies the absence of particle–hole symmetry as well—two criteria for topological SC.

As indicated in Fig. 3.6c, an increasing B_{ext} enhances the E_z and as a fraction of the E_T magnifies the region of ϕ over which the topological SC phase may be stable. The rationale here is that the single-electron character (characterized by diffusive transport and related to the E_T) is being magnified over the Cooper pair aspect

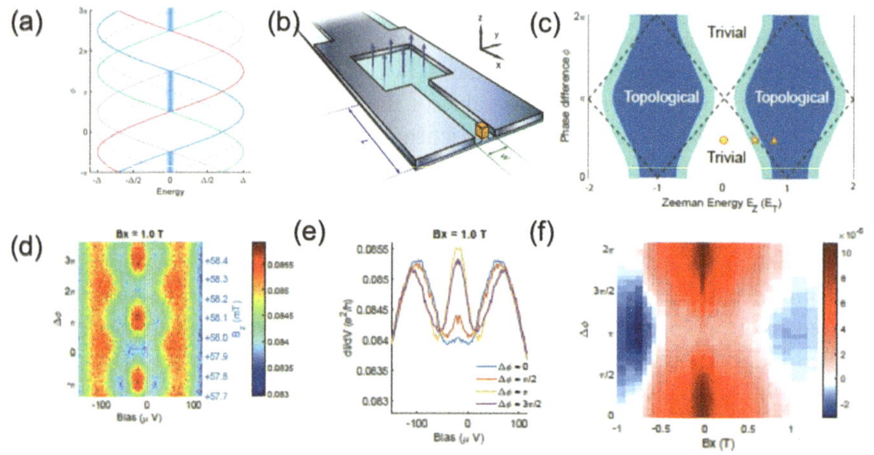

Fig. 3.6 *Topological SC in a Josephson junction (JJ) for zero-bias peaks (ZBPs) related to a Majorana-bound states (MBS).* (Adapted from [95]) (**a**) The influence of the phase (ϕ) tuning as a function of the SC energy gap. In the presence (/absence) of the B_{ext}, the related energy states in the gap are degenerate—*solid lines* (/spin-split—*dotted lines*). The *shaded region* close to zero energy represents the range of ϕ over which the topological phase develops, with corresponding ZBPs. (**b**) A schematic of the experimental apparatus, in a JJ configuration, formed by Al leads—with a junction length (L) of 1 μm and width (w) of 600 nm at one end and closed at the other end to form a *flux loop* The Ti/Au electrode—orange cube, functions as a *tunnel probe,* and is separated from the underlying HgTe 2-DEG by a layer of CdHgTe. A combined AC + DC bias is applied between the probe and the SC Al lead. The DC bias varies the energy of the topological SC phase corresponding to the horizontal axis of (**a**). (**c**) An increasing B_{ext} $(/E_z)$ as a fraction of the E_T magnifies the region of ϕ over which the topological SC phase may be stable up to a limit. (**d**) An experimental comparison to (a), at a particular B_{ext} ($= B_x$ of 1 T, in the figure) and $\boldsymbol{B}_{\perp}$ ($= B_z$ of the order of ~1 mT, as on the right hand axis), as a function of the DC bias. The differential conductance (in units of e^2/h) is on the right of the figure. (**e**) A line cut from (**d**) at various values of the ϕ, indicates a ZBP that corresponds to the onset a topological SC phase. (**f**) The zero-bias curvature of the conductance (G'') the magnitude of which is indicated on the right hand side) as a function of B_{ext} ($= B_x$), where a negative curvature (region marked white) corresponds to the ZBP

typical to the trivial SC phase. However, further increase of the B_{ext} yields more order to the diffusive aspect and destroys the SC as well, which is indicated through the reduced region of the topological phase.

An external Ti/Au electrode—indicated in orange in Fig. 3.6b and separated by CdHgTe from the 2-DEG in the HgTe and functioned as a *tunnel probe*. A combined DC voltage bias related to the energy variation on the horizontal axis in Fig. 3.6a was applied, and the AC current between this electrode and the SC lead (using an additional small AC voltage bias in addition to the DC bias) was monitored to obtain the two-terminal differential conductance ($G = dI/dV$). Plots of the G a function of the $B_{\perp}$ ($= B_z$) as well as B_{ext} ($= B_x$) yielded respective oscillatory variations: Fig. 3.6d. A typical tuning range of the former was ~1 mT while B_{ext} was varied up to 1 T: Of similar order of magnitude to the earlier study in Fig. 3.5. A minimum B_{ext} (> 0.6 T) is necessary to overcome reflection at the normal–SC interface. The lobes of enhanced conductance at positive (/negative) bias are related to electron (/hole) currents and the magnitudes and spatial extent are related to the particle–hole symmetry. The corresponding line cuts at various values of the ϕ are indicated in Fig. 3.6e. A pronounced ZBP was observed, with a maximum G of the order of 0.08 e^2/h, at ϕ differences corresponding to $E_z = ½\, E_T$—see Fig. 3.6c, somewhat related to the onset of the topological SC phase. It was indicated, through numerical studies, that a ZBP with the most contrast occurs at the onset of the topological phase, with possibly higher magnitude ZBPs deeper into the phase region. The delocalization of the Majorana wavefunction was inferred from such considerations. From an experimental point of view, as in the earlier study, such an aspect was considered through a negative value of the G'' as the relevant metric for evidencing the onset of the topological SC phase, and its variation as a function of the B_{ext} (= B_x) was plotted in Fig. 3.6f. Negative values were observed in the range of > |0.6| T, with correspondence to what was observed in Fig. 3.6d. In contrast to the previous work, which relied on the third-harmonic component of the measured current for obtaining the G'', here, the curvature was extracted through fitting a parabola to the measured G.

In this study, there was no clear indication of the possible formation of the Majorana modes due to plausible delocalization, for example, the magnitude of the ZBP was of the order of 0.03 G_o. The length scales of the channel, for example, w, dictate the typical scale for the manifestation of the Majorana state. For instance, a w of ~600 nm and an effective electron mass of ~0.03 m_e imply an induced gap of the order of 5 μeV. A more localized state may be obtained through a smaller w as well as a very long L. The ZBP may again be ascribed, from Fig. 3.5a, as related to transport from the *tunnel probe* into one of the ends of the channel in the JJ, where the Majorana state may be situated. As the energy broadening of the ZBP is an order of larger magnitude, the peak is clearly manifested and corresponds to an increased *DOS* where the system passes through a zero-energy level at the mid-gap. The subgap conductance was thought to be dominated by the ABS as well as a reflection-related broadening of the energy levels, along the entire length of the junction, with minimal contribution from the Majorana-related states given their delocalized aspect. However, when reflection is weak, then the ABS-related energy bands are flat with minimal dispersion and a related high *DOS* [95] which would again contribute to increased conductance as well as the ZBP magnitude.

Simulations of the channel supported the experimental observations qualitatively. Strict numerical agreement seems generally difficult due to the inherent assumptions that need to be made, for example, semi-infinite SC leads and normal region, etc., needed for the establishment of the Majorana state/MBS. Moreover, issues related to instrumental error of the order of 20 μeV as well as corrective issues related to magnet alignment may also preclude clear conclusion.

In summary, phase-related measurements on the existence of the MBS, through the DC Josephson effect have not yielded concrete evidence. The magnitude of the ZBP has not been clearly indicated to be G_o, and the invasive nature of the probing as well as the finite size of the flux loops used for the interrogation do not seem to yield proof of the existence of the MBS. We will next investigate the likelihood of the AC Josephson effect to evidence the MBS.

As indicated earlier, a trivial SC may be distinguished from a topological SC in the *I*–*V* characteristics, through illumination with EM waves (of frequency: *f*). While for a trivial SC, there would be characteristic voltage steps at values of $V_{n,SC} = n\,(h/2e)\,f$, with *n* as an integer; for the topological case, we would have $V_{n,T\text{-}SC} = n\,(h/e)\,f$—corresponding to the single electron nature of the related excitations. The transition from the trivial *to* the topological phase would be related to the disappearance of the odd Shapiro steps.

3.2.2 Response of a Josephson Junction to AC Radiation (External): Missing Shapiro Steps

An instance of an experiment related to investigating the fractional AC Josephson effect was carried out in Nb SC/InSb NW/Nb SC-based JJs probing the topological SC induced in the InSb by the SC Nb [90]. A DC SQUID–based configuration was used where the interference between the currents through *two* parallel JJs constituted from (i) Nb–weak link–Nb, and (ii) the InSb topological SC–weak link–InSb topological SC was studied: Fig. 3.7a. Such a SC–weak link–SC configuration, is often invoked as a textbook prototype for the Josephson effects, and the weak link was arranged in the related study through gaps of length in the range of 20 nm–120 nm. The ratio of the respective currents in the two junctions was estimated to be ~10, indicating one of the two JJs dominate. The current flow was then considered to occur primarily through the topological SC-based JJ and would be related to the phase difference particular to such junctions. The configuration also corresponded to a resistively shunted JJ [32], characterized by an $I_C R_N$ product of 1 mV (> $2\Delta_{SC}$) indicating a clean limit, where the coherence length (ξ) as well as the mean free path was larger than the weak link length—considered to be a "proper condition" for one electron tunnelling inherent to the possibility of observing a Majorana mode.

The JJ was subject to *external* microwave excitation in the radio frequency (*rf*) range of ~3 GHz with a related magnitude (V_{rf}) in the range of 0–10 mV. The resonant excitation of the Cooper pairs (in the trivial SC) or the single electron states (in the topological SC) are manifested as the Shapiro steps in the *I*–*V* characteristics [20]. It was shown that at smaller B_{ext} (= *B* in Fig. 3.7b—applied along the InSb nanowire, that is, perpendicular to the B_{SO}) Shapiro steps corresponding to Cooper pair tunnelling were observed, while at larger B_{ext}, > 2 T a doubling of the first Shapiro step was observed, that is, the absence of the first odd step, was evident: Fig. 3.7b. Here, the applied *B* presumably yields the required $2E_Z$ (of ~12 meV) and access to the related mid-gap states with an applied rf voltage (V_{rf}) of ~6 mV. The fusion of the Majorana particles to yield a *single electron*, and normal electron–related tunnelling–related signature, was posited to infer related topological phenomena. However, it was unclear from the presented data whether any other *odd* step was absent, as would be expected for the topological phase. Moreover, the conversion of the V_{rf} to actual applied voltage/energy onto the SC, that is, given an

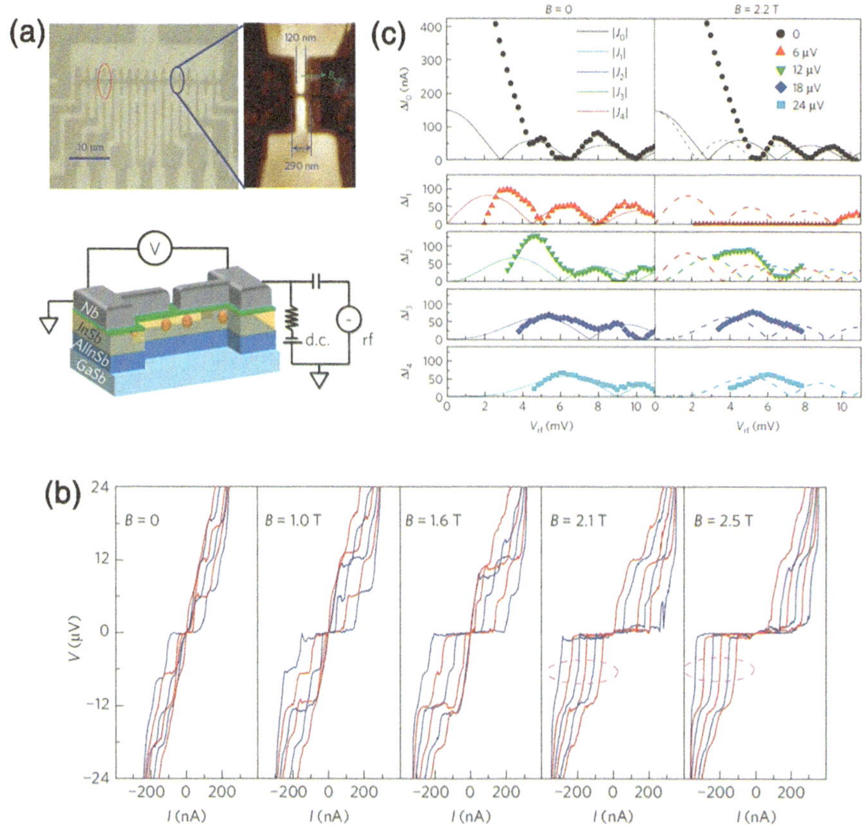

Fig. 3.7 *Identification of an MBS through an inverse AC Josephson effect.* (Adapted from [90]) (**a**) Top left, optical image of a sample with several Josephson junction (JJ) devices. The red oval outlines a single DC SQUID device comprised of two parallel JJs, with an enlarged AFM image of a single JJ on the right. The light areas are Nb. The green arrow indicates the direction of the spin–orbit field B_{so}. A schematic view of the device is at the bottom, and shows the possible location of the Majorana modes: orange dots in the topological InSb SC layer. (**b**) I–V characteristics of a JJ with increasing magnetic fields ($B_{ext} = B$). At B_{ext} less than 2 T, the Shapiro steps corresponding to the resonant frequency (f_{res})-based excitation of the SC states with the EM (at radio frequency: *rf*) waves can be observed. However above 2 T, the first Shapiro step (at a $V \sim -6$ μV) disappears. The dashed ellipse presents the possible conditions related to the presence of the MBS. (**c**) Plot of the width of the first five Shapiro steps (at a $V = 0$, 6 μV, 12 μV, 18 μV, and 24 μV, respectively) as a function of the EM field amplitude V_{rf}. Lines are the expected Bessel function–based fits of the form: $A|J_n(\beta V_{rf})|$, with $\beta = (2e/hf_{res}) = 0.84$ mV^{-1} (solid) and 1.04 mV^{-1} (dashed). At $B_{ext} < 2$ (*left*), the widths follow a Bessel function variation, with A = 150 nA. At a $B_{ext} > 2$ T (on the *right*), the width of the first step is zero and remains so over a range of V_{rf}. Here, the combination of the parameters related to B_{ext}, V_{rf} have been tuned to obtain the MBS. The lack of a proper fit with the Bessel functions seems to suggest that there could be a mixing of different frequencies, considering the nonlinear characteristics of the JJ

expected attenuation of three orders of magnitude and related energy scales pertinent to the device was unclear. It was also intriguing to note the result that the odd step "reappears" using larger rf voltages, which may indicate wave mixing effects as contributing to the missing step, as seen in Fig. 3.7c. The possibility of mixing between modes inside and outside the gap was considered.

The width of the constant voltage Shapiro current steps was monitored and fit to a Bessel function—as may be expected from the sinusoidal variation of the phase [$\delta \sim \frac{2eV_n}{\hbar\omega}\sin(\omega_n t)$]in the Josephson current relation: $I=I_c \sin(\delta)$. While the amplitude of the Bessel function is expected to decay for Cooper pair quasiparticle related transport, a relative constancy of the amplitude was used to distinguish Majorana-related electron transport. As cautioned, the Shapiro step evolution must properly be considered through *both* an applied current bias as well as from an applied voltage bias points of view [20]. It was indicated, for instance, that careful understanding of the external circuit resistance as well as the *I*–*V* characteristics in the limit of the DC Josephson effect, would be necessary. Indeed, the evolution of the current plateau widths, in the Nb/InSb case, did not fit Bessel function descriptions and the discrepancy was ascribed to contributions from "oscillations of different frequencies": Fig. 3.7c. Consequently, the implications of the phase matching of the first odd step to incident rf powers seems to be a matter of concern and needs to be explored more clearly, considering the postulated occurrence of the Majorana-related states exactly at the mid-gap. Finally, higher magnetic fields (> 3 T) resulted in the disappearance of all observed Shapiro steps and was related to the suppression of the excess current from the mid-gap states, as well as Andreev reflection–related effects. Broadly, the energy scales in the experiment are unclear with respect to the magnitude of the $2E_Z$ gap, as considered through the applied rf voltages. Moreover, the contributions of the single-electron-related current transport from the Majorana particle fusion–related current seems not lucid.

In another representative study, the absence of the first Shapiro step, while the other odd steps were present, was again attributed to the 4π-phase variation in the JJ constituted from HgTe as the topological SC induced from a overlaid Nb [91]. A mean free path of 200 nm seems to indicate a clean limit to the JJ. Here, a suppression of the $n = 1$ step was again observed at low-excitation frequencies (~2.7 GHz corresponding to an energy of 11 μeV), while the step reappeared at higher frequencies with gradual diminution as the frequency is reduced. Here, an energy metric would be related to the Josephson frequency ($f_J = \frac{2e}{h}V$), which translates to ~483.6 GHz/mV. For a V related to the $I_C R_N$ product (~1 mV), the resonant frequency of 2.7 GHz is not readily rationalized. A speculation is that larger time scales (corresponding to lower frequencies) allow for possibly necessary energy relaxation processes in the topological SC. The gradual fading away of the step as a function of applied rf power, at a fixed frequency (<11.2 GHz) indicated such a threshold effect as well. It was also observed that the $n = 1$ step is absent at lower rf currents and is subject to oscillations at higher rf currents; such aspects, sensitivity to external fields and forces, etc. all throw some doubt on the plausibility of the missing peak being considered a robust signature of Majorana states. Indeed, "trivial ballistic states," for example, as related to the abundance of modes, was posited by the authors as plausible for the observed experimental observations. The possibilities of Landau–Zener (L-Z) tunneling related to the diabatic transitions at the energy crossing points, which now occur due to the 4π periodicity, as well as to a quasi-particle continuum [96], are other factors that may set a time scale. The related energy scale of ~10 μeV estimated from the tunnelling probability are in some accord with the frequency dependency. However, this may imply that the $n = 1$ peak may not be present at higher frequencies. It was also discussed that thermal poisoning driven by Joule overheating may provide a reason for the resilience of the higher odd Shapiro steps [97].

In summary, it is unclear whether a clear signature of the MBS states, harnessing the AC Josephson effect, was obtained through external microwave excitation. As indicated earlier, the magnitude of the related RF waves at the position of the device is unclear. Such lack of control precludes a clear correlation of the *resonant* influence of the EM waves and their role in probing MBS-related single-electron states. Moving the EM wave source closer to the JJ may enable better control. We next describe an *on-chip* design where the excitation source is related to NW-based JJs, with related Josephson radiation produced by application of appropriate bias voltages. A SC tunnel junction is employed as the detector, through which the missing Shapiro steps may be inferred.

3.2.3 *Response of a Josephson Junction to AC Radiation* (**Internal**)*: Single Electron Conduction*

Generally, when radiation is incident on a JJ, in addition to the resonant excitation of the supercurrents in the SC, which results in the Shapiro steps, there is also a photon-assisted tunneling (PAT) component related to the quasi-particle (i.e., single electron) current through the SC. Enhancements of the currents, in the latter case, occur at photon energies that correspond to the Δ_{SC} and related voltages: $V_{PAT} = n\ hf/(e)$, again at a voltage twice that corresponding to the Shapiro steps. Consequently, frequency sensitive measurements aimed at uncovering the response of a NW-based JJ were performed [96], and monitored through a detector biased at V_{DET}. It was noted that the f is not related to the Josephson frequency (f_J) but is more dependent on the junction dynamics, that is, related to the energy scales of the L–Z tunneling between the branches of the junction bound state and the quasiparticle continuum. For the experiments, the f range was in the order of 1–20 GHz. The radiation from the NW JJ, considered as an oscillator, and whose frequency was tuned through voltage bias, is coupled capacitively to a detector based on two JJs constituted from $Al/AlO_x/Al$ tunnel junctions arranged in a SQUID-based configuration. For a fixed bias across the NW (V_{NW}), a relation: $hf = e^*V_{NW} = 2\Delta_{DET} - eV_{DET}$ holds, where $e^* = 2e$ (for conventional SC, with Cooper pairs) *or* $e^* = e$ (for topological SCs, involving single-electron tunneling characteristic of Majorana modes). The physical process related to the detection corresponds is related to the PAT. Lock-in techniques used to excite the NW yield subsequently a frequency dependent current, sensed in the detector: Fig. 3.8b, c. The transition from the trivial to nontrivial regimes was done through tuning the B_{ext} as well as the chemical potential of the NWs, using external electrodes. Now, depending on whether the NW in the trivial SC or nontrivial SC regime, the monitoring of the differential conductance (dI_{PAT}/dV_{NW}) would yield information on e^*—see Fig. 3.8d, e. A maximum in the (dI_{PAT}/dV_{NW}) was related to the tunneling of the e^*. A transition of e^* from $2e$ to e was interpreted to be due to the change of periodicity from 2π to 4π periodic Josephson effect: Fig. 3.8f. The notion that different charge states, corresponding to e^*, could be excited in a SC, depending on its topological state and related to different frequencies is an interesting idea.

However, it was unclear whether the assumption of a constant voltage spectral density: $S_V(f)$ was accurate, with respect to the fidelity of transcribing the recorded I_{PAT}. The present experiment also does not shed light on the frequency dependence of the behavior of the $n = 1$ peak observed previously [91]. Moreover, the reliance on internal JJ dynamics would imply the necessity of calibrating each individual device. The aspect of experimentally determining e^*, while subject to interpretation—see Fig. 3.8d, e, was also related to the fractionalization in anyons and could be further investigated.

In summary, it is unclear whether modulations of the *I*–*V* curves or the conductance, through the Josephson effect, could be considered as adequate evidence for states which could host the Majorana modes/MBS. For instance, relaxation oscillations, due to inductive effects in JJs [98], have been proposed to yield related period doubling and sub-harmonic effects. Additionally, the relevant absence of the odd Shapiro step, for providing evidence of the 4π periodic effect, has often been probed at low currents. The possibility of transitioning to chaos and related periodic doubling bifurcations have also been invoked in resistively shunted Josephson tunnel junctions [99]. It is then plausible that variations in the JJ parameter space, due to the use of specific materials may confuse the interpretation. Moreover, the expected [100] admixture of the *s-wave* and the *p*- or *d-wave* character, would mean that the measured current could involve contributions from *both* the conventional and unconventional AC Josephson effects. It was also considered as to how non-topological effects related to the hybridization of the Majorana modes, for example, in shorter nanowires involving say, partially separated Andreev-bound states, and SC gap suppression may give rise to 4π periodicity [101, 102]. In summary, the large number of conflating effects and the perturbation of several energy states in the EM excitation confuses the obtaining of a clear signature of the MBS, in both the DC and AC Josephson effects.

It would be next considered whether the MBS may be directly probed, say by tunneling directly into the states, through scanning tunneling microscopy (STM)–based methods. While such studies may not directly help with the practical utilization of the MBS in a circuit-based configuration, probing the likelihood of the related modes even existing would be of scientific importance. We will discuss a few related studies related to exploring zero-energy modes in lower dimensions. A seminal study was related to probing for the Majorana states at the end of iron atom chains.

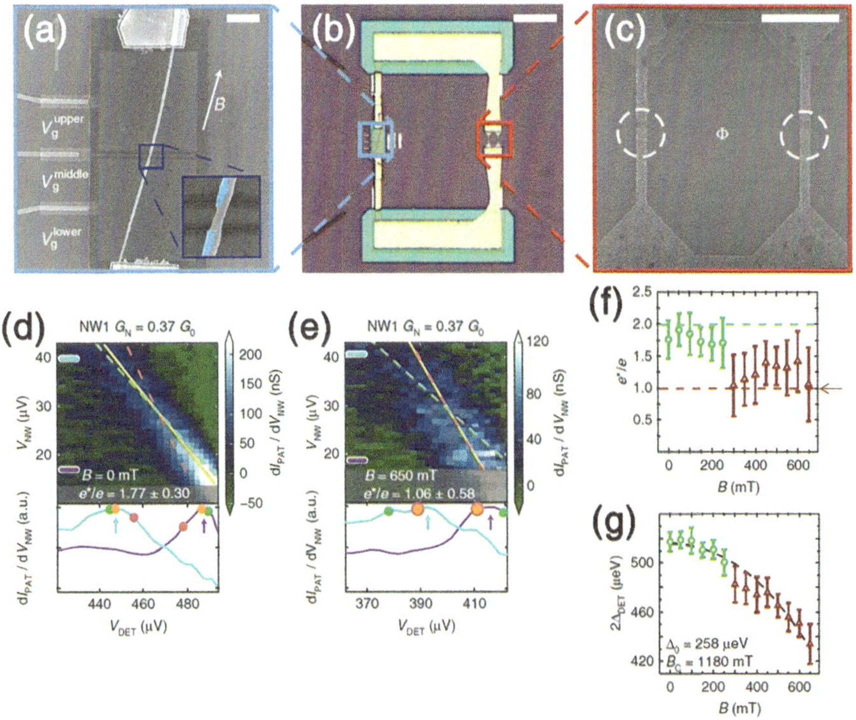

Fig. 3.8 *Inferring MBS from the response to internal Josephson junction radiation.* (Adapted from [96]) (**a**) Micrograph of the SC Al–covered InSb NW-based Josephson junction (JJ). The NW was placed on three electrostatic gates to control the tunnel barrier and chemical potential (μ) of the NWs, so as to tune the *trivial* and the *nontrivial* SC phases. The inset shows a false color micrograph of the junction with epitaxial Al, grown on the NW, for inducing SC. Scale bar: 1 μm, (**b**) An image of the coupled NW junction (*left*) with a detector (*right*). Scale bar: 10 μm, (**c**) The detector is comprised of Al/AlO$_x$/Al JJs (in dashed circles) connected in a SQUID configuration, Scale bar: 1 μm. (**d**) & (**e**) The differential conductance maps related to the photon-assisted tunneling (PAT) current as a function of the voltage across the nanowire (V_{NW}) and the voltage across the detector (V_{DET}) at 0 mT and at 650 mT, respectively. The orange line represents the fit $e^*V_{NW} = 2\Delta_{DET} - eV_{DET}$. The green dashed line corresponds to $e^* = 2e$ (corresponding to Cooper pairs in a trivial SC) and the red dashed line corresponds to $e^* = e$ (corresponding to transport by single electrons as in a topological SC). The cyan and the purple lines at the bottom correspond to the horizontal marked lines in the top, with the orange, green, and red dots indicating the position of the *best fit*, as well as the $e^* = 2e$ and $e^* = e$, fits respectively. The arrows correspond to the peaks of the differential conductance, signaling tunneling through a single energy level that may correspond to the Majorana mode. (**f**) The variation of the (e^*/e) with externally applied magnetic field (B_{ext}/B) indicates a transition from the trivial to the nontrivial SC regimes at ~300 mT. (**g**) A plot of the SC gap width for the Al on the detector side (Δ_{DET}) as a function of B, with the dotted line fit to: $\Delta_{DET} = 2\Delta_0\sqrt{1 - B^2/B_c^2}$, related to the nominal decrease of the SC gap with increased B, as would be expected for a trivial SC

3.3 Majorana Modes in Ferromagnetic Atomic Chains

An interesting notion related to the discovery of Majorana modes was experimentally investigated through considering the related interaction between ferromagnetism (FM) and superconductivity. At the outset, such a study is intriguing considering the typically understood mutual incompatibility between the two phenomena, that is, an aligned spin state is considered in variance with the two spin state assumed for the Cooper pairs in traditional *s*-wave SC [102]. Moreover, the itinerant character of the electrons for FM character would be reduced in chains of finite length. However, it was indicated that proximal SC could still be induced in the Fe atoms, with a related bandgap and a demonstrable transition from a trivial SC phase to a topological SC insulator phase. The presence of S–O coupling would further lift the spin degeneracy and in the presence of a B_{ext} render the topological phase effectively spinless—in line with the principles discussed in Fig. 2.1. The resulting modulation of the spin arrangement in the atomic chains leading to a *spin texture* could be mapped to a topological configuration.

Consequently, one-dimensional FM wires constituted from Fe atom chains were proposed as a platform for identifying Majorana modes in the topological SC phase [104]. The idea of using atoms is, at the very outset, relevant in that a better spatial resolution may be obtained, for example, through atomic-level STM, in contrast to NWs where there has been an ambiguity in the effective separation of the putative Majorana modes to a distance much shorter than the geometric length.

Topological phases in magnet atom chains could also take the form of helical texture through the mutual interaction of the constituent spins. However, such spin chains [105] arise from magnetic moment coupling and are susceptible to intervening disorder, which could confuse the evidence for the Majorana modes. It was then considered to use well-aligned moments, that is, in the FM or anti-ferromagnetic (AFM) coordination, and subsequently deploy S–O coupling with an underlying SC, which would introduce a SC gap. In comparison to earlier studies using NWs, the Zeeman field would be provided by the Fe itself and compete with the introduced SC gap. As seen earlier (Sect. 2.1), such *competition* is necessary for a trivial *to* nontrivial phase transition. A relevant gap could then incorporate zero-energy states compatible with the Majorana modes, which could be introduced at the ends of the atom chains.

Nadj-Perge et al. [104] probed a system of closely spaced Fe atoms placed on top of a Pb substrate—an *s*-wave SC (T_c ~ 7.2 K, and with a gap $2\Delta_{SC}$ ~ 2.73 meV) with intrinsic S–O coupling, and proximity-induced SC in the Fe atom spin chain: Fig. 3.9a, b. The Fe atom chains, ~20 nm–50 nm in length, were synthesized through in situ sputtering of Fe on the Pb surface. Deposited Fe islands were implicated in nucleating the chains, few atomic layers of which were thought to be submerged into the Pb substrate.

A substantial charge transfer from the Pb to the Fe was found through density functional theory (DFT) calculations. The magnetic interactions along with the S–O interactions were then investigated to determine the conditions necessary for the formation of the topological SC phase. Firstly, the exchange interactions between the closely spaced Fe atoms were shown, through band structure calculations, to yield fully occupied majority spin bands and partially occupied minority spin bands—with the implication that the E_F lies within the latter. It was observed that on varying the chemical potential, an odd number of bands crossed the E_F, laying the basis for a topological configuration, with an odd number of edge states, as encapsulated through the Z_2 classification [19]. A *p*-wave related gap of ~0.1 meV was observed as indicated in Fig. 3.9c, along with the possibility of Majorana modes for

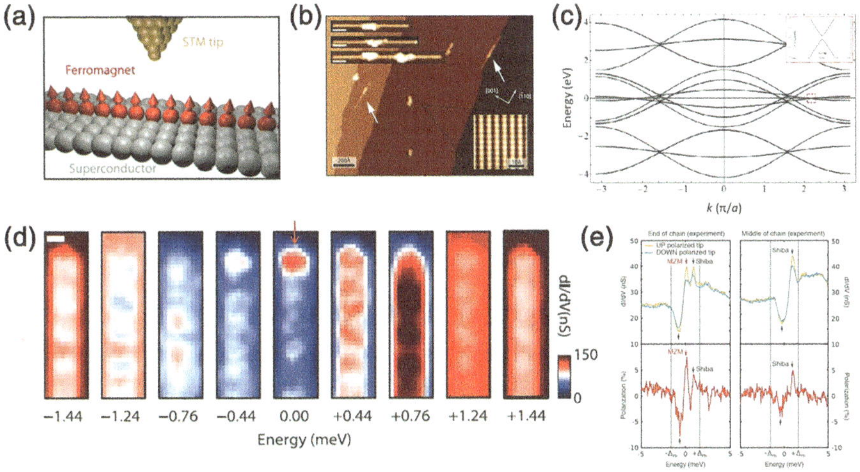

Fig. 3.9 *Majorana modes in atom chains*. (Adapted from [104]) (**a**) A schematic of scanning tunneling microscopy (STM)-related probing of Fe atom chains (tuned to a topological SC phase) embedded in an *s*-wave superconductor (Pb). (**b**) Topographic image of the Pb (110) surface with Fe islands and chains grown on top (shown by white arrows). (*Bottom inset*) Atomic structure of the Pb(110) surface. (*Top inset*) Images of the Fe chains and nucleating islands. The white scale bars are 5 nm. (**c**) Calculated eigen energies computed from the Bogoliubov de Gennes (BdG) Hamiltonian for the straight Fe chain proximal to the *s*-wave superconductor. The inset indicates a zoomed-in image of the red area in the main figure indicating the presence of an effective *p*-wave SC gap of about 1 meV. (**d**) Spectroscopic and spatial differential conductance mapping along the atom chains. The bias voltages/related energies on the STM tip are indicated at the bottom. At zero bias, there is observed enhanced conductance corresponding to a localized Majorana mode assigned at the end of the chain. The red arrow specifies the signature corresponding to the Majorana mode. Scale bar is 1 nm. (**e**) (Adapted from [117]) Spin-resolved STM measurement using a paramagnetic tip may be used to resolve trivial states, for example, as induced from magnetic impurities, from the nontrivial states, that is, the Majorana modes. The spectral variation of the differential conductance at the end (*left*) and the middle (*right*) of the Fe atom chain. The ZBP related to the Majorana modes (at $E = 0$) and the Shiba modes (at $E \neq 0$) are both observed at the end of the chain, while only the asymmetric Shiba modes are observed in the middle. The figures at the bottom indicate the spectral variation of the polarization (defined through the ratio: $[G_P(E)–G_{AP}(E)]/[G_P(E) + G_{AP}(E)]$, where $G_P(E)$ ($/G_{AP}(E)$) is the measured differential conductance with the tip magnetization parallel (/antiparallel)—corresponding to the UP (/DOWN) polarized tip configuration/s. A nonzero polarization at zero energy corresponding to the Majorana mode states (red arrows) is seen at the end of the chain whereas, a near zero polarization is observed (at zero energy in the middle of the chain), with sub-gap associated Shiba states (black arrows)

a one-dimensional system. Such computations involve the solution of the BdG Hamiltonian [106], wherefrom a Majorana number may be computed as a function of chemical potential and the exchange interaction, and describes a topological phase, for example, with two Majorana modes at the two ends.

A spatial resolution of the Majorana modes was sought to be indicated through scanning tunneling microscopy (STM), with both normal and SC tips. The STM study was stated to have a resolution of ~0.1–0.2 meV. In (i) the constant current mode, the height of the tip may be varied to keep the tunneling current constant, obtaining a topographical image of the sample. The spatial variation of the local density of states (*LDOS*) at low temperature and with relatively energy-independent tunneling matrix elements [56] could also be calculated from this data. While in (ii) the constant height mode the tip distance from the sample is kept constant and at each position in the plane, the tunneling current is measured for varying voltage bias, yielding the differential conductance which is again proportional to the *LDOS* in the sample. The spatial and spectral conductance mapping could be used for estimating the *LDOS* as a function of energy at any point in the sample, yielding insights into the SC gap as well as sub-gap structure [107]. The use of a SC tip in the STM would yield even greater resolution, even more than the thermal limit, due to the convolution of two sharp *DOS* features intrinsic to the tip and the sample SC, respectively. Here, the coherence peaks of the sample SC appear at a voltage (V) given by $eV = \pm (\Delta_{tip} + \Delta_{sample})$ and concomitantly sub-gap states (with energy E) are resolved with $eV = \pm (\Delta_{tip} + \mathrm{E})$. Alternately, a zero-energy state corresponding to the putative Majorana mode would correspond to resonances with $eV = \pm (\Delta_{tip})$[109]. Here, it is assumed that the tunneling rate of electrons from the tip is slower than the quasiparticle relaxation in the SC.

Differential conductance mapping, as a function of the tip (in the normal state) bias, revealed increased values toward the end of the chain, as in the zero-bias-related signals in the middle of Fig. 3.9d, and the localized character of the mode. Such an aspect was considered the *main experimental feature* related to the confirmation of the Majorana mode/quasi-particle in this and other studies [108, 109]. The extent of the localization of the mode was related to the ξ in the SC with respect to both the SC proximity gap as well as a helical field gap of the chain, the latter related to the formation of spin texture along the FM atom chain [109]. While the first localization length could be larger than the chain length, the latter length is more representative of the spatial extent of the mode. For instance, the respective localization lengths were estimated for a ~ 6 nm Fe chain to be of the order of ~22 nm (~60 atom sites) and ~ 0.7 nm (~2 atom sites), respectively. Moreover, short localization lengths are expected due to strong velocity renormalization in the hybrid system [110, 111]. Then, the experimentally observed localization length would typically be much shorter than the ξ of a bulk SC. It was also observed that [112] only some of the chains showed the zero bias peak, along with an associated possibility that the ZBP could be overlapping the coherence peak of the induced topological SC gap, which was of the order of 0.1 eV. However, there is a pronounced delocalization of the mode at higher bias/energies and also evident spectroscopic features in the Pb SC gap (at <1.36 meV). The mapping of the atomic location to the mode itself does not seem very clear.

Measurement on the Fe atom chains with SC tips (e.g., Pb-coated W tip) yielded a pronounced enhancement of the *dI/dV* to ~60 nS at an energy corresponding to 1.36 meV (the SC gap of Pb). Weak coupling between the STM tip and the Majorana mode at the end of the atom chain was invoked for a *G* orders of magnitude below the expected G_o (of ~77.5 μS). The coupling strength between the tip and the mode was smaller in magnitude compared to the thermal energy (corresponding to ~0.12 meV at 1.4 K). Interestingly, at a point in the middle/bulk of the atom chain a *dI/dV* peak of magnitude ~90 nS at 1.66 meV was interpreted as due to the *p*-wave gap (of magnitude 0.3 meV, cf. the theoretically predicted gap of 0.1 meV), while another peak was noted with $G \sim 38$ nS. It is also relevant to note that the Majorana mode–related resonance is regarded symmetric while alternate sub-gap resonances are asymmetric, from the underlying uneven contributions from the contributing electron and hole states. Moreover, the line-shape of such resonances would also depend on their distances from the ends of the chain. Consequently, a related confusion with respect to a proper delineation of a Majorana mode could be due to its possible overlap with such sub-gap resonances.

3.3.1 Possible Artefacts in the Measurements

The aspect of obtaining Majorana quasi-particle modes (MQPMs) [104] hosted by magnetic atoms coupled to a nominal *s*-wave SC does not seem to be universal but may instead be related to details of the appropriate chemical potential as well as proper correspondence to band structure calculations. For instance, an exploratory study considering Co atomic chains (with lengths ranging from 2.5 nm to 11.7 nm) proximity coupled to a Pb substrate did not indicate any clear evidence of Majorana modes [108] but instead revealed a zero-bias peak-related signal delocalized along the whole length of the chain. Local confinement effects as well as variations in the local potential were proffered as possible reasons for a finite zero-energy spectral weight along the chain. One plausible reason for the absence of topological character was speculated to be due to an even number of Fermi energy crossing points in the Brillouin zone, which hybridize and prevent the formation of a topological phase related to Co atoms. Alternately, Fe, with one less electron, has three Fermi energy crossing points, with a related odd crossing of the SC related E_F yielding the possibility of a topological phase. Consequently, yet another criterion for the formation of Majorana modes is the consideration that the spin direction/related magnetization in the atom chains be perpendicular to the length of the chain, fulfilling the necessary spinless attribute.

Possible artefacts that could contribute to the observed ZBP were also sought to be ruled out. For instance, in the presence of magnetic impurities, the Kondo effect related to a modulation of the conductance may arise. It was considered that Kondo-related resistance peaks, typically due to magnetic impurities, would persist beyond the SC–normal transition temperature, and was ruled out through experiments carried out with Pb in the non-SC regime. Another criticism was that the SC gap induced in the FM chain is of the order of 100 μeV, implying a very large coherence length (ξ) compared to the length of the atomic chains. As indicated earlier, the localization length of the wavefunction for the Majorana modes is related to the ξ. However, in the experiment it was observed that such a localization length was only around 10% of the length of the chain. Moreover, all the experiments were performed at ~1 K, where quasiparticles could be thermally excited to unoccupied sub-gap states. Such excited quasiparticles could conceivably poison the Majorana mode and diminish the energy resolution as well as the intensity of the ZBP compared to G_0. To counter this, a further study at lower temperatures (250 mK) was carried out which showed further evidence for Majorana states [113]/MQPMs. There was no detectable splitting of the zero-bias peak, and the peak was particle hole symmetric for some of the chains. It was also observed that the peak had a substantial spectral weight in the substrate.

However, SC yet seems to be essential to the formation of the putative Majorana modes, as such features were not observed when the SC was turned off, for example, through an external magnetic field. Local disorder as well as related potential variations are generally noninfluential for inducing sub-gap states in *s*-wave type superconductors. Such aspects are in accordance with Anderson's theorem [56] which asserts the robustness of the SC phenomena. However, there may still be magnetic impurities which induce such sub-gap states [57].

3.3.2 Possible Contributions of Magnetic Impurities and Their Resolution

The violation of time reversal symmetry, in and around magnetic impurities, would be expected to suppress *s*-wave superconductivity and the related SC bandgap. Moreover, magnetic atoms coupled to a SC have been shown to induce in-gap states, for example, the Yu–Shiba–Rusinov (YSR)/Shiba states [58, 114, 115]: Fig. 2.4b. Such states may be spin-polarized to an extent that is determined by the exchange coupling of the magnetic impurity spin with the electrons in the SC [57]. The sign of the exchange interaction parameter (J) is also relevant, in that the influence of the Kondo screening (which is often associated with anti-ferromagnetic exchange) on the SC gap must also be considered. For a FM coupling of the impurity spin to the conduction electrons, the related bound state is typically close to the edge of the SC gap, inducing relatively weak carrier scattering. The related wavefunctions of the bound states are localized over the magnetic impurity states, with a $1/r$ decay characteristic. Increasing impurity concentration would also result in bands, with a bandwidth proportional to $\sqrt{n_{imp}}$ and contribute to a reduction of the SC gap, ultimately to a zero value. Moreover, while bound states close to the SC gap edge would imply weak scattering, strong magnetic scattering would result from *tail* states from the localized impurity band which may reside deep inside the gap.

The interaction of Fe atoms with the Pb-based bulk superconductor could also be modeled as related to the influence of magnetic impurities, which would introduce additional states into the SC gap [56] and modify the *DOS*. Such an effect is in contrast with the influence of small amounts of nonmagnetic impurities, where conventional superconductivity (with Cooper pairs that correspond to pairing of time-reversed quasiparticle states yielding Kramers degeneracy of the states) is protected—from the Anderson theorem. However, it was also discussed that with a *unconventional* SC [56] constituted from a higher orbital angular momentum state, such as a *d*-wave SC, or even perhaps a *p-wave* superconductor that any scalar/nonmagnetic impurity and related scattering would result in pair breaking and suppress the SC phase. The impurity related resonances are generally not expected to be at zero-energy if explicit particle–hole symmetry is absent. There was some evidence of such magnetic impurity scattering–related states through the observation of Shiba bands in the SC gap.

Majorana states may be considered a subset of such YSR states present in topological superconductors. As the YSR states are described as unequal superposition of an electron and a hole operator, on changing the polarization direction of the tip, the response from the two energetically symmetrical YSR states will be different. Both extended and localized Shiba states could exist anywhere along a chain of atoms, while the Majorana modes would only be derived from extended electronic states and be ideally present at the very end of the chains. While the YSR states correspond to scattering-based resonances, for example, as related to bound states within the SC gap and have a finite energy relative to the gap center, there is still the possibility that the local potentials could accidentally perturb a YSR state to reside at the zero-energy level. The distinction between such accidental states, such as the *trivial* YSR states and a Majorana zero mode could be made through spin-polarized spectroscopic STM measurements [116].

Considering the possible influence of magnetic impurities, the normal states' spin-dependent contribution was compensated by a *set-point* effect [117]. Here, a magnetically polarized tip probes the differences between up and down spins through, for instance, height changes and is calibrated off through recording the related tunneling current change. The measurements of the energy-dependent spin polarization: *P*(E), reckoned through the difference of the tunneling current conductance: δG (E), between when the tip is polarized parallel and antiparallel to the normal to the Pb substrate surface, may be used as a metric to distinguish trivial modes from the nontrivial Majorana zero modes. The measurements of the polarization were carried out of the atomic chains over an energy window of ± 5 meV around the expected E_F, initially when the Pb is in the normal state, to set the zero-scale of the polarization. The SC is then turned on and the spin polarization remeasured over the length of the chain, to look for spin contrast in concert with the differential conductance mapping. The tip magnetization itself was only expected to influence the sign of the spin contrast and not its magnitude.

It was indicated, through such measurements, that the end of the atomic chain exhibited enhanced spin contrast with a differential conductance larger, by a factor of two at ~35 nS, compared to regions away from the chain end. Polarization features related to in-gap states, such as those that may be attributed to the YSR states, were also observed in the tunneling spectra. However, an important aspect as determined for the YSR states was an antisymmetric/odd character of the conductance change, that is, $\delta G(\mathrm{E}) = -\delta G\ (-\mathrm{E})$, implying an equivalent positive increase of the G on one side of E and a negative decrease of the G on the other side with a *net* value close to a zero value at zero bias. Consequently, the P(E) of the YSR state should be zero at zero energy as well. Here, the polarization contribution of the trivial YSR state was similar in the normal as well as the SC state, and may also be subtracted out through the *set point* methodologies indicated earlier. Alternately, the observation of a spin contrast indicates the presence of a nontrivial mode, such as the Majorana state. The related spin polarization arises from the minority spin bands that intersect with the E_F.

The charge transfer from the Pb to the Fe, and related dipole-like asymmetry may be rationalized as giving rise to the asymmetric tunneling. The antisymmetric spin component would also imply a symmetric spatial aspect to the impurity-related wavefunction. However, such contrast was observed for the purported Majorana mode/MQPM at the ends of the chain. As it was considered previously that the E_F intersects with the minority spin bands, such spin related texture could be justified. Given that the Majorana mode is a feature of an extended topological band structure, the relevant background would also be delocalized. As indicated earlier, the calibrating set point effect related to the STM experiments, in the presence of an external magnetic field of ~100 mT to suppress the superconductivity from the Pb substrate and the Fe, compensates for spin-related background contributions from the extended states. The difference in the related Hilbert spaces, that is, local for impurity-induced magnetic/nonmagnetic states versus nonlocal/extended for the Majorana modes is yet another distinction enabling such spin contrast to be manifested at a specific point at the end of the chain. A minimum-length scale for the Fe atom chains, at least larger than 3–4 nm, was found to be necessary for the observation of the Majorana modes and was argued as precluding end disorder effects.

As previously indicated, both the conductance change and the difference in the conductance change, as indicated by the P(E), may be used for a clear distinction between the trivial and nontrivial states. It is indicated that trivial localized states would not yield a P (E) at an energy value of zero, corresponding to the mid-gap. For instance, even if a YSR state could be tuned to zero energy, the resultant P(E) would be less than that contributed by the normal state background. If it is assumed that the set-point method has been successful in eliminating the background contrast, ruling out the possibility of alternate states, it could then be inferred that a state which would yield a significant P (E = 0) would indeed correspond to a Majorana mode.

It was then indicated, through STM measurements [117], that a peak in the polarization spectra of the modes was observed at zero bias at the ends of the chain, for example, Fig. 3.9e, and consequently attributed to be related to the *localized* Majorana mode. This mode is discriminated from the *delocalized* magnetic impurity–related YSR/Shiba states through the finite value of the P (E). The YSR/Shiba states are (i) present at both the end and middle of the chain, (ii) exhibits an antisymmetric conductance profile, and (iii) has a *net* polarization at zero bias was approximately zero (interpolating between the Shiba state resonance).

In summary, the study related to identifying Majorana modes/MQPMs in Fe atom chains ascribes a spin-related contrast for their identification. Alternate magnetic impurity states in the SC gap were discriminated through a lack of spin contrast. Such an attribute arises from the aspect that a STM tip polarization could suitably alter the spin state of the impurities, while having no such effect on the MQPM. There has then been ascribed a nonlocal or *delocalized* as well as a *localized* character to the minority spin bands of the Fe; with the former related to the normal state background and the latter to the MQPM.

Considering such caveats, and allowing for the possibility of localized MQPMs, it is still unclear as to how such modes as well as the Fe chains can be incorporated in a practical device, that is, the interaction of the formed modes in zero-dimensional atomic-scale systems would still need to be carried out in a three-dimensional device manifold. It would, perhaps, be judicious to harness the characteristic end/corner states of a one-dimensional system integrated in a three-dimensional material. We will see that such harness is possible through considering higher order topology, exploring the zero-dimensional to zero-dimensional hierarchy. In this context, a better understanding of the protected states in a topological insulator (TI) would be of utility. Moreover, the underlying S–O interaction could be further enhanced where the related interaction could be essentially infinite, that is, through specific confinement of states in the corners of spinful edge channels at the periphery of the TI (see Sect. 2.1).

3.3.3 Majorana Modes at Topological Insulators (TI)–Related Interfaces

A fundamental property of a *d*-dimensional TI is the existence of lower-dimensional gapless states at the boundary [118]. Ordinary TIs possess d^{-1}-dimensional gapless states. Recently discovered higher-order TIs, that is, HOTIs, such as Bi, possess related states of dimension d^{-2} or lower [119–123]. These gapless states (hinge states) are protected by certain global symmetries, for example, time reversal symmetry (TRS), inversion symmetry, or chiral symmetry (comprised of particle–hole symmetry and TRS)—inducing immunity of particle-like zero-dimensional states to magnetic forces [124]. Each dimension reducing step may be characterized by its own global symmetry. The gapless states may be manipulated by explicitly breaking symmetries. For example, the TRS may be broken by applying magnetic fields or polarized optical pumping.

The interface between the related one-dimensional helical *hinge* modes and Fe atom chains—again on top of a SC substrate (Nb) were posited to host the Majorana modes [125]. The related study was complementary to much of the earlier work related to the identification of the Majorana modes conducted in topological SCs *synthesized* through proximity-induced superconductivity in (a) materials systems with a large S–O coupling and related internal magnetic fields for inducing spinless/*p-wave* character, and involving (b) Tis, where the intrinsic gap is modulated by the SC gap, so as to induce band gap closing and reopening and concomitant zero-energy states, at the crossover. As one essence of a TI is the presence of edge modes, it was sought to investigate whether such states may host the Majorana modes [125]. A dimensional reduction from one (characteristic of the edge states) to zero (for the point like Majorana modes) is implicit.

There is then again the possibility of inducing SC in such topologically protected edge and related lower-dimensional states. Such states exhibit ballistic transport and may be probed to yield quantized conductance (G_o). As TRS is preserved, the edge states intrinsically manifest extremely large ("essentially infinite" [125]) values of an intrinsic B_{SO}. The zero-energy states related to the Majorana modes may be identified at the intersection of edge states or through the specific interaction of edge states, say with Fe clusters. Such an aspect was probed in [125] where Bi hinge states were placed in proximity of a Nb-based SC to create one-dimensional topological SC systems. As indicated in Sect. 2.1, the ingredients, related to (i) the breaking of spin symmetry through an internal magnetic field from the Fe clusters, and (ii) the particle–hole symmetry, are both achieved at the interface. Here, 3D Bismuth as a HOTI would be the host of the topologically protected hinge states [126].

The Bi films were synthesized through deposition on a cooled Nb single crystal (110) surface. The one-dimensional hinge modes are manifested in Bi bilayer constituted islands on top of a Bi (111) thin film surface. The bilayer Bi (with trivial gapped surface states but topologically protected gapless hinge states) embodies topologically protected edge states on every other edge of the hexagon in the bilayer, classified as an A edge: Fig. 3.10a, depending on whether the surface atoms are proximate to the vacuum [127]. Here, a one-dimensional *DOS* characteristic of an edge state is indicated. Alternately, the set of Bi atoms closer to the surface would be of the trivial kind and exhibit expected characteristics, for example, a two-dimensional *DOS*—Fig. 3.10b.

Consequently, scattering free one-dimensional propagation would be feasible along the A edges, with enhanced S–O coupling as the basis. Further, Fe was deposited on the Bi bilayer, and was predominantly clustered, for example, as islands, at the corner of the A and B edges of the hexagonal Bi as indicated in Fig. 3.11a. The topological SC is now induced into the Bi Type A hinge modes, from the SC Nb, and the Majorana zero modes (MZMs) are posited to exist at the end of the mode. The coupling of the isolated helical Bi edge modes to the Fe cluster energy levels yields a magnetic hybridization gap (Δ_H), as related to induced magnetism in the Bi from the Fe and would be larger than the related Zeeman gap (Δ_M): Fig. 3.11b. The larger Δ_H extends through much of the Brillouin zone around the E_F, with minimal requirement on the control of the chemical potential through carrier density modulation, as in the NW studies, *cf.* Figure 3.1 or Fig. 3.2. Instead, the hinge–Fe particle interface represents a point where both the SC gap (Δ_{SC}) as well as the Δ_H both tend to zero, yielding the requisite zero-energy state. It was indicated, through STM

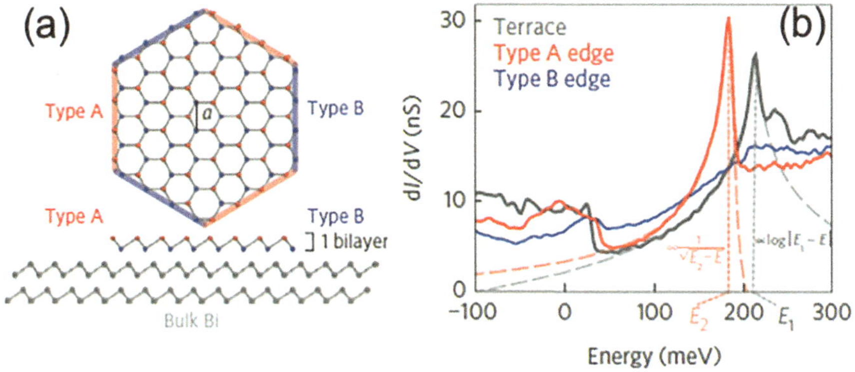

Fig. 3.10 (**a**) The proximity of Bi atoms on the surface to the vacuum or the substrate yield the Type A and Type B edge states, respectively. (**b**) A 1-D DOS attribute is related to the Type A edges, as examined through STM spectroscopy, while the Type B edges exhibit typical 2D-DOS behavior. Such differences form the basis of higher-order topology in Bi. (From [127])

studies, that the magnetization of the Fe cluster should be along the hinge mode direction for obtaining maximal Δ_H.

As indicated earlier, the spin-contrast (present in the Majorana modes and absent in the Shiba states) of observed zero-energy states, may be indicated through spin-polarized STM spectroscopy and used as a diagnostic for the MZM, distinguishing them from the trivial modes. STM measurements were done using a superparamagnetic tip (Fe-coated Cr) on the Fe clusters to find their magnetization directions as inferred from the tip magnetization. Further STM measurements were done along the edge from the center to the Fe cluster: Fig. 3.11c, and a Majorana mode was observed at the interface: Fig. 3.11d. As an illustration of the significance of the Type A edge, it was indicated that such ZBP features were not observed with Fe clusters on the Type B edges.

The spatial location of the Majorana modes were then inferred through the ZBPs observed through STM around and at on the cluster–hinge interfaces on the A edges as indicated in Fig. 3.11d. The manifestation of the MZM mode and a Shiba mode is evident at the edge and center of the Fe cluster, with a slight asymmetry of the latter. A somewhat clearer delineation is indicated in Fig. 3.11e, with plots of both the differential conductance (*top*) and the polarization (*bottom*) of the center (*left*) and the interface (*right*). The signal from both up and down magnetized tips are shown, where the contrast is much smaller for the Shiba states compared to the Majorana zero modes (MZM), at close to zero-bias. It was noted, with the given assignment of the tip magnetization, that the measured polarization of the state could only be positive for the MZM while it could be of either sign for the Shiba state. The magnitude of the differential conductance of the Majorana mode, halves at around 2 nm away from the interface. While such a finding was in accord with the theoretical expectation, the latter considers a fitting through the renormalization of

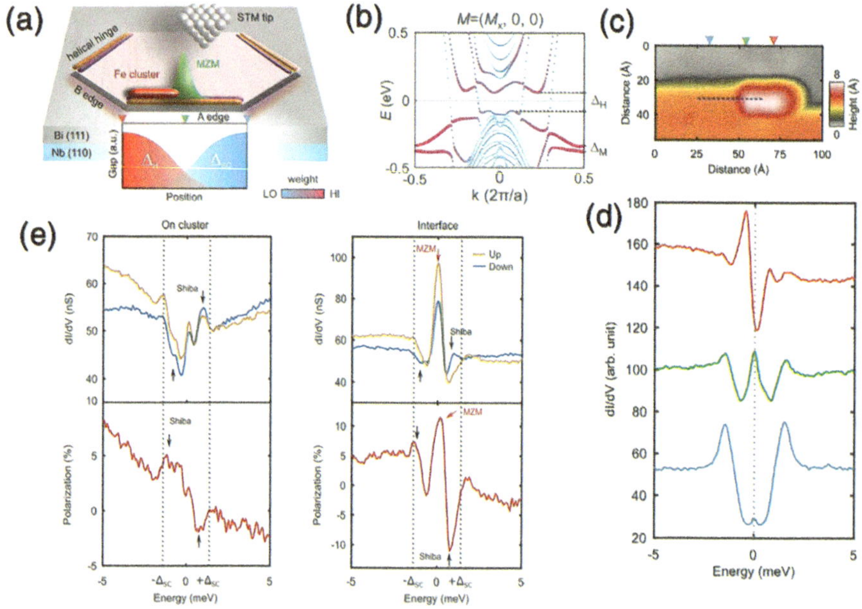

Fig. 3.11 *Majorana modes in higher-order topological insulators (HOTIs).* (Adapted from [125]): (**a**) Materials arrangement for the hosting and STM observation of a Majorana zero mode (MZM). The higher-order topological insulator (HOTI) characteristic of Bi in which topological SC is induced in specific edge states, that is, Type A by the Nb, is invoked. The MZMs (shown in green) have been assigned to form at the intersection of the helical edge/hinge states on Bi bilayers and Fe atom clusters. (**b**) The band structure of the edge covered by the Fe cluster. Due to hybridization between the magnetic cluster and the helical hinge modes, a magnetic hybridization energy gap (Δ_H) throughout the Brillouin zone at the Fermi energy. The Δ_M is the Zeeman gap, if Fe cluster wouldn't have been present. (**c**) A topographic image along the edge, showing the Fe cluster (red oval). Local conductance maps were taken along the dotted line. The gray background is the Bi (111) layer and the orange layer is the Bi bilayer. (**d**) Three conductance maps at the three different locations marked by arrows of the corresponding color in (**c**) in the different regions along the edge cluster. The red map describes a FM region associated with the Fe atoms, the green curve at the interface shows the ZBP in the SC gap and the blue curve taken in the middle of the edge shows a small ZBP characteristic of the tail of the Majorana mode at the interface. (**e**) The spin resolved STM measurement on the interface and the Fe cluster. The top two figures show the conductance maps with up-polarized and down-polarized tips. The bottom two figures show the energy dependence of the polarization: *P* (E). A non-zero polarization seen at zero energy is characteristic of the MZM at the interface. The black arrows show the in-gap Shiba states whereas the red arrows indicate the MZM

the edge mode Fermi velocity from the proximal SC, which again imbues uncertainty on the topological state and its degree of immunity from the surrounding.

The existence of the Majorana mode at the interface on the A edge has been due to the lack of hybridization with the bulk or surface state at that edge, while such coupling does occur along the B edge. The SC gap at the A edge is then from the surface state as well as the topological edge state, with the latter fitted through BCS theory. The measured gap width (through differential conductance) of the topological state: Δ_T, was quite large and the same as that of the underlying Nb superconductor gap width: Δ_{SC}, that is, of the order of ~1.5 meV. Considering that the BCS theory is typical to *s*-wave SCs, such an assignment related to the topological gap is not obvious. The aspect that the magnetization of the clusters need be parallel to the edge, for inducing the Δ_H in the topological edge state, and the related formation of the Majorana modes is interesting in light of Sect. 3.3.1, where it was posited that the spin direction in the atom chains would need to be perpendicular to the length of the chain. Consequently, the character of the topological state in such systems needs to be further investigated. The location of the other Majorana mode, tentatively assigned to the interface of the Fe cluster-B edge, is also ambiguous in these investigations. While a *highly delocalized* nature has been assigned to the alternate mode arising from the coupling of the B edge states to the bulk modes, the combination of the two modes that would be integral to a Majorana bound state is confusing. The usually ascribed topological protection does not seem to be apparent in such disparately formed and connected modes.

In addition to the possible elimination of the contribution of magnetic impurity-related states to the observation of the MZM, any Kondo effect–related contribution was ruled out through the control application of a magnetic field which eliminates the SC. Here, the ZBP was absent indicating that the induced SC was essential.

A noteworthy aspect of the work is the utility of a domain wall as a manifestation of a topological feature, and related topological invariant, for the system. Here, the wall represents an interface, in energy space, between a FM-related bandgap and a SC-related bandgap. The related topological change, between two *spinless* phases necessarily passes through a zero-energy state attributed to the Majorana mode. The polarization interaction, between the Bi edge state and the Fe cluster, was modelled to be of the RKKY(Ruderman–Kittel–Kasuya–Yosida) type [44], manifesting an oscillatory variation of the sinusoidal type. Moreover, the aspect that a magnetic field is not necessary, for the opening and closing of the band gap as was done in NW-based architectures, is appealing. The presence of other magnetization components, say from impurities, reduces the Δ_H, enabling the hybridization of the Majorana modes with such states with a loss of non-Abelian character. Consequently, changing the magnetization in the ferromagnetic clusters can be used to switch on and off a single Majorana zero mode. However, as considered earlier, the effect on the other Majorana mode through such modulation is unclear along with the related implementation in quantum information processing. Speculatively, the delocalized Majorana mode could perhaps be utilized for perhaps coupling the localized mode to the continuum.

Moreover, the presented scheme may be extended to Bi nanowires, as indicated by the authors, where the Majorana modes are pinned to the ends and may subsequently invoked in specific Majorana pairing Schemes [49]. The notion of Fe atoms as marking the Majorana quasi-particles is attractive for delineating their respective location. An advantage of using Bi is the related small effective mass and large carrier mean free path, whereby the quantum confinement effects can be observed at much larger length scales. Bi nanowires are easier to fabricate as Bi has a very low melting point (~271 °C), which makes the synthesis amenable to melt-injection and vapor-deposition techniques. For instance, nanowires with diameters around 7 nm have been fabricated in porous alumina (electrical insulator) templates, which can be easily prepared with good control over pore size, via electrolytic processes [128]. There is also a critical wire diameter (d_c) at which there is a transition from semimetal to a semiconductor state [129], which may be used for further tuning the character of topological states.

In summary, the notion of reducing the degrees of freedom (i.e., through avoiding chemical potential modulations) to invoke Majorana mode formation is appealing at the outset. However, (a) the lack of information on the location of both the modes, (b) the seemingly necessary precise control involving Fe cluster-A edge interfaces, and (c) the lack of information on the dynamic characteristic of the topological edge state into which the SC as well as the magnetization is induced are all confusing factors that would hinder practical implementation. Finally, the recording of the conductance significantly below the G_o presents another issue in pinning down the mode characteristic.

Such experiments and ideas on topological insulators invoke the use of 1-D edges and their ends, which are essentially 0-D in character, for locating Majorana states. We will next explore whether edge states can be considered manifestations of the Majorana modes, as probed through the quantum anomalous Hall effect. Subsequently, the observations on the ends of such 1-D edges, as may be found in the vortices of SCs will be studied. The coupling of the 0-D Majorana modes with such lower-dimensional features of bulk materials would enable their integration into a device form.

3.4 Majorana Modes in Quantum Spin Hall/Quantum Anomalous Hall Insulators

The spinless aspect of Majorana modes, as related to the edge states, may be observed through the transport as considered in terms of the Quantum Hall effect (QHE), where an *external* applied magnetic field induces edge states [11, 130] The QHE does not resolve the spins and consequently may then be mapped to Majorana modes. Indeed, early proposals indicated the possibility of finding Majorana quasiparticles in specific quantum Hall states [131]. However, the necessary particle–hole symmetry intrinsic to SC is diminished or even destroyed in large magnetic fields, typical necessary for the manifestation of the QHE. Alternately, the transport channels in a materials system manifesting the quantum anomalous hall effect (QAHE) are guided through *internal* magnetic fields, produced for instance, through spin–orbit (S–O) coupling. It has been indicated, for instance, that such magnetic insulators coupled with topological aspects of the band structure would be *both* necessary for manifestation of the QAHE [132].

3.4.1 *The Quantum Anomalous Hall Effect (QAHE) and Chiral Majorana Modes*

The significant feature of the QAHE then is the quantization of the transverse/Hall conductivity to e^2/h, in the absence of an external magnetic field and when the E_F is situated in the middle of the insulator gap, along with a vanishing of the longitudinal conductivity [133].

Inducing SC in a topological material, manifesting the QAHE, would then be expected to yield particle–hole symmetric states in the material. Such states were termed as Chiral Majorana edge modes (CMEMs) and occur in pairs (corresponding to unresolved up and down spins) on each edge [134]. These modes are to be considered based on coherent charge transport and fulfil the criteria of delocalization related to the Majorana modes. A particular CMEM can be isolated through the application of a small external magnetic field, of the order of 0.1 T, which would suppress the transport of one spin type through the topological SC [15, 135, 137], but which is nevertheless small enough so as to not to perturb the SC state (where the relevant magnetic fields would need to be of an order of magnitude larger). The time reversal symmetry relevant to the QAHE is also broken through such nominal fields. Indeed, transitions in the band structure, yielding a range of electronic phases, for example, from a trivial insulator—trivial SC regime *to* a *nontrivial* SC—topological SC regime, may be induced, and yield a specific CMEM. Subsequently, in a range of magnetic fields, where the topological SC sandwiched between two QAHE materials transports only one CMEM, the longitudinal conductance would be quantized to *half* multiples of e^2/h due to the contribution from the CMEM. There is also the implication of perfect transmission of the CMEM through the topological SC. The directionality of the edge modes depends on the magnetization and the related exchange field induced from the dopants.

The related 1D Majorana modes were posited to follow conventional fermionic exchange statistics [138]. It was also posited that the propagation of the chiral modes leads to similar unitary transformations as the braiding (see Sect. 1.3) of the Majorana zero modes [139]. The dynamic aspect can then, in principle, would then be a step ahead of just isolating the modes at the end of NWs, and subsequently devising methodologies for their exchange. Topological insulators, such as Bi_2Se_3 with bonding and anti-bonding *p*-orbitals as the basis, doped with FM transition metals have been predicted to exhibit the QAHE [132, 140]. An insulating magnetic state is enabled through dopants with the 3+ valence, in addition to ensuring a zero *DOS* due to the dopant, say through a gap between the majority and the minority spin bands. Fe and Cr were found to be suitable dopants. Here, the E_F would be in the gap, and could subsequently be controlled through an external gate voltage to yield a minimum (/quantized values) of the longitudinal (/transverse or Hall) conductivity, characteristic of a quantum Hall effect. At zero external bias, a careful control of the film thickness is essential to maintain the two-dimensional character and the related coupling between the top and bottom surface states in the film. The latter characteristic determines a trivial gap (which may be of the order of ~4 meV

[141]) which may be overturned, into a nontrivial gap, through an internal exchange field from the S–O coupling, perpendicular to the film surface, characteristic of the anomalous Hall effect. Edge states may then propagate with energy states in the *nontrivial* gap region.

3.4.2 Possible Manifestation of Chiral Majorana Modes Through Experiments on Coupled SC–Quantum Anomalous Hall Insulator (QAHI) Systems

Some evidence for conductance quantization, that may be relevant to the presence of Majorana modes was seen in experiments performed on a magnetic topological insulator in the QAHI compound: $(Cr_{0.12}Bi_{0.26}Sb_{0.62})_2Te_3$ synthesized on a GaAs (111) substrate by molecular beam epitaxy (MBE) [141]. A Nb superconductor was proximity coupled to the TI and the resultant conductance through the QAHE manifested through a magnetic TI-topological SC–magnetic TI arrangement, yielded such evidence: Fig. 3.12.

The thickness was controlled to be six quintuple layers (to a total of ~6 nm), considering the aspect of coupling between the top and bottom discussed earlier and Nb (~200 nm) deposited in the central part of the MTI with contacting Au electrodes: Fig. 3.12a. Two-terminal *I–V* measurements of the conductivity of the topological SC (with the Nb strip) and four-terminal measurements on a control QAHI (without the overlaid Nb) were done. An additional back gate is deployed at the bottom to tune the topological properties of the QAHI. When Nb is in the SC state, topological SC is induced into the QAHI, and can be subsequently modulated through an external magnetic field (*B* in Fig. 3.12). Here, the *B* tunes the magnitude of the exchange field, and governs the trivial to phase changes in the QAHI as well as in the TSC. The *B* was typically of the order of 0.3 T, and necessary to overcome the coercive field of the QAHI as well and is smaller, by an order of magnitude, compared to the critical field of the Nb SC. The effects of the tuning of the *B* are indicated in Fig. 3.12b. The spinful channels are developed in the presence of a saturating *B* and yield the dual edge modes encompassing both the QAHI as well as the TSC region, as indicated in (i) and (vii). Case (i) (/(vii)) corresponds to a saturation of the magnetization in the $+z$ ($/-z$) direction and subsequent reduction in the $-z$ ($+z$) sense, with opposite sweep directions of the *B*. The evolution is then from (i) → (ii) → (iii) for *B* field sweep in one direction, and (vii) → (vi) → (v), for a *B* sweep in the other direction—at the end of which a single CMEM is trapped. In the intermediate states, (ii) and (vi), obtained at reduced *B* from that corresponding to complete magnetization saturation, there is only the possibility of a single CMEM in the TSC. The QAHI is still stable and propagates two modes, one of which is reflected at the QAHI–TSC interface. It is noted that states (iii) and (v), correspond to a further reduced *B* field regime where there is now a transition from the conducting QAHI state to a trivial/normal insulator (NI) state, while the SC is still in the topological phase—implying that the CMEM could effectively be trapped in the

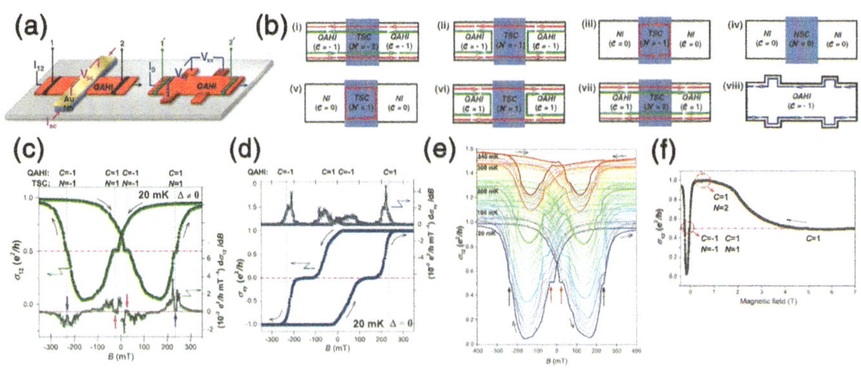

Fig. 3.12 *Chiral Majorana edge modes (CMEM).* (Adapted from [192]): (**a**) A schematic of the experimental design invoking the production and filtering of CMEMs through a hybrid quantum anomalous Hall effect insulator (QAHI)–superconductor (SC) device. The device on the *left* includes a thin film of the QAHI with a Nb SC strip in the center. Electrical conductance (σ_{12}) measurements along the length of the device were done through measuring the ratio of the voltage between terminals 1 and 2, while passing a current I_{12} between the terminals. The Nb SC characteristics, related to the electrical resistance, were monitored by the I_{SC} and the voltage across the SC (V_{SC}) recorded. The device on the *right* is a *control* QAHI sample, absent the SC, characterized through its longitudinal (V_{xx}) and transverse (V_{xy}) voltages, in response to a bias current through the sample. Additionally, the voltage difference across terminals 1′ and 2′ yields a longitudinal conductance $\sigma_{1'2'}$. (**b**) The edge transport configurations, related to the CMEMs under a sweep of an external magnetic field (B). The Chern number (C), related to the number of edge channels/modes, of the QAHI varies between − 1, 0, 1 whereas the C of the SC varies from −2 to 2. The B starts at full saturation—along an axis perpendicular to the sample, say + z, where both the constituent CMEMs belonging to a single chiral edge mode pass through the topological SC (TSC)—(i). On decreasing the B toward zero, say from such an initial saturation in the + z direction, there is initially a transition into one CMEM passing through the SC (ii) and on further decrease the QAHI transitions to a normal insulator (*NI*) phase—(iii), finally followed by a TSC phase transition to a trivial SC phase—(iv). Decreasing the magnitude of B (along the − z axis) follows a reversed pattern where B saturation yields the state indicated in (vii) followed by a single CMEM transmitting phase—(vi), QAHI-NI transition—(v) and a TSC to a trivial SC phase change—(iv). The mode propagation in the topologically nontrivial QAHI phase, in the absence of a SC, is indicated in (viii). (c) The variation of the σ_{12} in the hybrid QAHI–SC device, shown on the *left* in (**a**), and (d) the Hall conductivity: σ_{xy} in the control QAHI device, shown on the *right* of (**a**), as a function of the B. Half integer conductance plateaus can be observed when the topological phase transitions occur through the measurement of the σ_{12}. The lower (/upper) inset plot for (**c**) and (**d**) indicate the derivative of conductance with B with the peaks and arrows on the plots showing the respective topological phase transitions. (**e**) Temperature dependence of the σ_{12} of the QAHI–SC device with 10 mK steps. The red and black arrows show the half integer quantum conductance (= ½ (e^2/h) plateaus where each trace (except the lowest) was offset for clarity. The contrast of the ½ (e^2/h) plateaus diminishes with an increased temperature, possibly due to thermally activated poisoning of the CMEM. (**f**) The variation of the σ_{12} with B, indicates both the e^2/h and ½ (e^2/h) related phenomena. At low B, the e^2/h corresponds to the conductance in the QAHI, while above a critical B ~ 4 T, the Nb SC is in the normal configuration, corresponding to the series connection of the two QAHI regions in series and a reduction of the conductivity to ½ (e^2/h)

TSC phase. The transition of the TSC to a nontrivial SC state (NSC), on further B decrease, and the lack of any edge state modes is depicted in (iv).

With respect to the electrical measurements, it would then be expected that the evolution from a single-edge state (constituted from two CMEM modes) to a single CMEM in (ii) and (vi) would result in a conductance decrease from (e^2/h) *to* 0.5 (e^2/h). The CMEM is related to a channel of transport and corresponds to a particle–hole symmetric state [134]. Two CMEMs are *topologically equivalent* to one QAHI-related state. Such aspects could be considered quantitatively through the assignment of a Chern number (C) which would, broadly be either $C = 0$, or $C = \pm 1$, corresponding to trivial/nontrivial phases of the QAHI, respectively. A particular chiral edge transport corresponding to $C = -1$, in the entire QAHI material, is indicated in (viii) of Fig. 3.12b. The selection of a particular odd C state only permits the corresponding CMEM to propagate through the TSC. Consequently, the CMEM—which is related to the Majorana mode in the TSC phase, would have a contribution to the transport/conductance of half that of the *net* chiral edge mode manifold. The QAHI is then considered to possess two CMEMs for a given edge state (with $C = \pm 1$), and hence a SC with an even Chern number would have the same characteristics as the QAHI. For B, where the TSC has an odd C and the QAHI has a nontrivial C, plateaus at ½ (e^2/h) would be observed.

The modulation of the conductance was probed experimentally at temperatures ranging from 20 mK to 350 mK, the results of which are shown in Fig. 3.12c–f. The variation of the two terminal conductivity (σ_{12}), as a function of the B, and illustrating the variation of the C is also seen in Fig. 3.12c, where at large B, the quenching of the SC in the Nb would be expected to yield chiral edge transport through the QAHI, as shown in (viii) of Fig. 3.12b. It is also indicated in Fig. 3.12c that small kinks/plateaus corresponding to the ½ (e^2/h) values were observed, as related to (ii) and (vi) of Fig. 3.12b. A higher B, in the positive or negative sense, leads to a saturation of the σ_{12} to e^2/h values, as may be inferred from (i) and (vii) of Fig. 3.12b, through the edge state propagation through the QAHI–SC interface. The complete reduction of the σ_{12} to a zero value, expected when the $B = 0$, was thought to be prevented due to electrostatic fluctuations over the sample. In the control QAHI sample, without the overlying Nb: Fig. 3.12d, phenomena related to the QAHE were indeed observed in the Hall conductivity (σ_{xy}) measurement and indicates intrinsic magnetic field related effects. Additionally, the temperature sensitivity of the σ_{12}, in Fig. 3.12e, provides a measure of the rather fragile characteristic of the quantization over an energy range as small as ~1 μeV (~10 mK). The possibility of artefacts contributing to the conductance jumps, for example, through jerky domain wall movement in magnetization changes—the Barkhausen effect[45], was considered to be stochastic, irreproducible and to have an effect much smaller than the observed conductance changes. Moreover, two QAHI in series would be expected to yield the half conductance quantum plateau, if the intervening Nb was just a connecting element, and not a TSC. This was indeed observed in the σ_{12} as shown in Fig. 3.12f, at fields beyond 4 T. However, lower fields enhance the σ_{12} ruling out just such a simple contact between the QAHI on either side. However, from Fig. 3.12c only B of the order of 0.3 T were involved in obtaining the ½ (e^2/h) plateaus, suggesting two different mechanisms—one at low B and another at high B for the observation of the plateau regions.

However, the surface conductance of a prototype TI would be expected[143] to be quantized in units of $(n + 1/2)\ e^2/h$, and such higher-order plateaus are expected to be present, which were not evident in the observations. While a low B for the manifestation of the QAHE is somewhat of an advantage, compared to previous studies related to semiconducting NWs—as in Sects. 3.1 and 3.2, where external fields of the order of 1 T were needed, the poorer resolution of the plateaus and the correlation to an extended region where the CMEMs occur is an issue. Moreover, the aspect of incorporating two diverse phenomena, that is, SC interacting with a topological state, yielding a truly topological SC effect, and manifested through electrical transport measurements is attractive. While the fundamental notions related to a trivial *to* nontrivial phase transition with accompanying band inversion are still pertinent, a firm understanding of the coupled QAHI–TSC state remains to be achieved.

3.4.3 The "Absence of Evidence" of the Chiral Majorana Modes in QAHI–SC Hybrids

Generally, the coupling between the SC and the QAHI must be high for inducing CMEM propagation and requires high-quality interfaces as well as samples. In earlier studies, for example, related to symmetry protected topological states in NWs, such coupling was sought to be avoided through proximity-induced SC in the states. There is yet a degree of overlap and aspects related to the transparency of the QAHI–SC interface [143] were invoked for proposing an "absence of evidence" for the chiral Majorana modes in related devices [144]. In this context, a ½ e^2/h plateau, previously ascribed to the CMEMs could also arise just from a good contact between the SC and chiral edge modes in an insulator exhibiting the QAH [143], and such a feature itself would then not be able to distinguish the existence of 1D chiral Majorana fermion modes.

For proof of the occurrence of half-quantization in other scenarios, a device consisting of an Nb strip (width ~ 20 um) over the width of the QAHI and an Nb finger (width ~ 200 nm) attached to it on the right was fabricated [144]. The QAHI was comprised of 3QL Cr doped $(Bi, Sb)_2Te_3$/5QL $(Bi, Sb)_2Te_3$/3QL Cr doped $(Bi, Sb)_2Te_3$. The large degree of contact transparency between the SC Nb and underlying QAHI was monitored through the comparison of the measured conductance with respect to *both* Andreev reflection (when the external field: H, is less than the critical field: H_c of SC Nb) as well as normal reflection (when $H > H_c$). It was subsequently determined that the two terminal G was ~ e^2/h even when the Nb was not SC (with fields > H_c) indicating the coupling of the Nb to the underlying QAH chiral modes and yielding the half-integer quantization following the geometry of the left side of Fig. 3.12a. Then, the *decoupling* of the Nb layer from the QAHI was proposed as a more likely cause for the observation of the ½ (e^2/h) features, and consequently the topological SC phase itself was deemed to be not necessary. In the regime of good coupling of the SC layer to the QAHI sample, the ½ (e^2/h) conductivity was observed whenever the magnetization was well aligned in the sample, in a large magnetic field range up to 0.5 T. Additionally, an experiment involving the use of multiple (m) Nb strips with an as arrangement indicated on the left side of Fig. 3.12a [145] indicated a decreasing σ_{12} varying through fractions, that is, $[1/(m + 1)]e^2/h$, indicating that the earlier observations were probably just indeed from a series resistance contribution of the $(m + 1)$ QAHI sections. Consequently, the original experiments and related explanations were placed in some doubt. Moreover, the observations related to Fig. 3.12f need to be yet understood. Alternative explanations which do not invoke Majorana physics ("non-Majorana mechanisms" [144]) were also published, for example, where it was proposed that half quantized conductance could arise through percolation of quantum Hall edges, which could take place before the onset of the topological phase in the SC or at temperatures even above those corresponding to the SC gap [145].

In conclusion, the validity of forming as well as detecting Majorana modes through CMEMs is still under scrutiny. From one point of view, the idea of a CMEM is analogous to a continuous tunneling path through a mid-gap zero-energy state in a SC, as considered earlier for the diagnostics of the Majorana modes in NW-based architectures. Additionally, the use of a QAHI is laudable for harnessing the internal degrees of freedom related to the magnetic field, much like the basis for topological insulators involving S–O coupling (as previously discussed in Sect. 2.1). The issue seems to be rather the proper conduct and interpretation of experiments related to the harness of the mobile CMEMs, invoking the Majorana modes, while preserving particle–hole symmetry. For the latter, a SC state seems necessary while tuning the SC to a topological phase to ensure the passage of one of the two CMEMs remains the issue. Considering the promise of using such a CMEM for practical quantum information processing [139], further related experimental investigations would need to be of high priority and interest.

3.5 Majorana Modes in the Vortices of Superconductors

It will be recalled that at a SC–TI interface, that proximal SC could be induced into an edge state at the TI surface. The consequent electronic state integrated with the SC may be corresponded to a vortex-like state in the latter, and a Majorana mode could form at the end of such a vortex. It would then make sense to interrogate the possible occurrence of topological states (TS) at the ends of the vortex cores in SCs [146]. Such an aspect is appealing in that it is well known that the application of a magnetic field onto a Type II SC results in an array of normal metal vortices inside the SC: Fig. 3.13a. It is briefly noted that a Type I SC differs from a Type II SC, mainly in how an applied magnetic field interacts with and penetrates into the SC [21]. The vortices span the thickness of the SC and represent magnetic flux tube penetration into a SC. The vortex size is related to the coherence length (ξ) which ranges from 1 nm (in high T_c superconductors, such as $YBa_2Cu_3O_{7-x}$) to hundreds of nm in conventional superconductors, such as Al [20]. The SC is parameterized through an order parameter, which approaches zero at the very center of the vortex. The parameter is *stiff* outside the vortex and accounts for the rigidity and stability of the SC state. Further, the spectroscopy of vortices and vortex cores has been observed and studied in detail. The vortex array is stabilized by a screening current enclosing exactly one flux quantum [20] per vortex. Interestingly, a zero-bias peak (ZBP) manifesting a finite conductance at zero-voltage was found in seminal experiments on probing the vortices—see Fig. 3.13a and determining a plausible connection to zero-energy states would then be of much appeal.

Consequently, the ends of a vortex core found in a SC were indicated to possess such unique zero-energy modes. Given that the manipulation of the magnetic flux in such vortices forms the basis of significant technology, for example, the flux shuttle [147] and the related single flux quantum (SFQ) logic-based family of devices [32, 148], it would be of significant technological interest to probe the vortices. The nature of the core may be probed through a scanning tunneling microscope (STM)–based study and the inherent states that are present may be identified [149]. The related spectroscopy of vortices and vortex cores has been conducted and studied in detail over the past few decades. Techniques such as single-particle (/quasiparticle: QP) tunneling spectroscopy and SC pair (Josephson) tunneling have indicated the excitation spectrum of the SCs, delineating the bound states, Andreev states, pair coherence and the quantum interference of the pair wavefunction. For instance, the excitations (related to the formation of quasiparticles) from the SC ground state are a quantum superposition of the electron and holes arising from the SC condensate of electron pairs (as described by the BCS theory [20]). Such quasiparticle excitations may yield TS and/or zero-energy states situated exactly in the center of the SC gap and would need to be carefully delineated.

Quasiparticle spectroscopy through STM/STS has allowed the probing of the vortex lattice: Fig. 3.13b, and, in more detail, the interior of a SC vortex: Fig. 3.13c, which indicates the spectral probing of the vortex core in $NbSe_2$ (a *s*-wave SC) [149]. The STM/STS-related SC pair (Josephson) tunneling could be used to probe, in more detail, the excitation spectrum of the SCs, delineating the bound states,

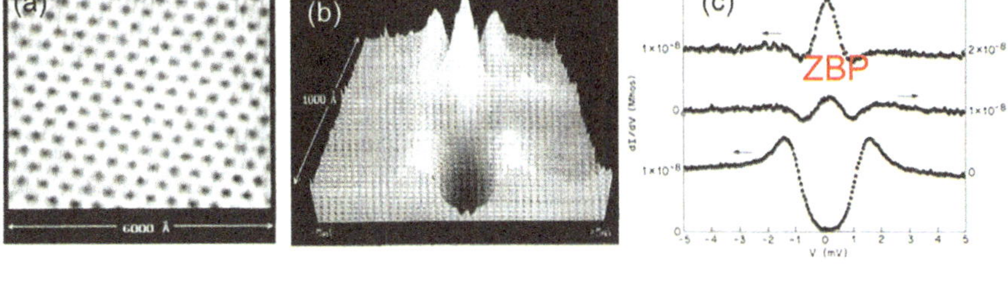

Fig. 3.13 (**a**) An Abrikosov flux lattice in $NbSe_2$ at 1.8 K, indicating isolation of the SC vortexes, of magnetic flux, (**b**) A mapping of the *DOS* inside a single fluxoid, (**c**) the differential conductance (*dI/dV*) vs. voltage (*V*) on a vortex (*top*), ~8 nm from the vortex (*middle*), and ~ 200 nm from the vortex (*bottom*). It is of much interest to investigate the correspondence of the ZBP (zero-bias peak) features [149] to zero-energy states and related *topological state* modes

Andreev states, pair coherence and the quantum interference of the pair wavefunction. Bound and well-defined QP states, of the Caroli–de Gennes–Matricon: CdGM type [150], were found in the center of the cores and related to near-perfect boundaries at the vortex walls of high-quality single-crystal $NbSe_2$. It was proposed that the ends of the SC may yield TS at zero-energy situated exactly in the center of the vortex [151]. The vortices may then be considered as regions inside which TSC is preserved. There is a strong suggestion that the related ZBP, which was found to have an energy width of 2 meV, may be further resolved into lower/zero-energy states perhaps corresponding to a TS. Previous work on probing the vortex core using STM on FeSe [152] and $NbSe_2$-related heterostructures [153] indicated the possibility of Majorana modes but lacked detail on resolution from the Andreev-bound states (ABS). Associated spin selective Andreev reflection [154] using spin polarized currents indicated a magnetic characteristic for the mode, while not explicitly considering the *DOS*. Moreover, possible geometric resonances were not carefully considered. Recent studies of the vortices of $FeSe_xTe_{1\text{-}x}$ (an *s*-wave SC) have indicated the necessity for investigating non *p*-wave SCs for the possibility of unique zero-energy states. A focus on the understanding and probing of the various possible modes and QP states in SCs, such as the ABS, as well as those due to geometrical resonances—in addition to the postulated zero-energy states relevant to Majorana zero modes would be helpful. Broadly, the BdG theory and related mean-field approaches [106] are quite suitable for a microscopic description of inhomogeneous systems, as in the case of vortices embedded in the background SC state. Modelling single and multiple zero modes and their interactions was demonstrated [155].

For the specific occurrence of the Majorana modes, additional symmetries as well as the spinless characteristic must be fulfilled, which could arise from time- and inversion-symmetric systems and an intrinsic spin–orbit coupling-related magnetic field (B_{SO}). Consequently, one may be able to work with SC with *inherent* topological behavior, with the advantage of reducing or even eliminating other constraints such as the necessity for clean interfaces and close coupling between SC and TI-based materials.

Consider first the SC–TI coupling, and related induced topological SC in the latter, as previously discussed (Sect. 2.1). The chemical potential (μ), related to the E_F, could be tuned in a range—from the middle of the TI band gap to deep inside the bulk SC bands. The former case would correspond to the formation of the zero-energy/Majorana modes, where with $\mu > \Delta_{SC}$, a zero-energy solution, is expected at the surface. When the μ is tuned to be in the bulk bands, low-energy excitations related to the CdGM-type states [150], bound to the core of a vortex line should be considered [156]. The eigenvalues of these excitations are separated by an energy spacing of the order of $\Delta_{SC}/k_F\xi$ (~ $\varnothing^2_{SC}/E_F$), where k_F is the Fermi wavevector corresponding to the SC gap: see Fig. 2.1.

There is then expected to be a critical μ (= μ_c) where there is a transition of the nature of the vortex from that corresponding to a nontrivial phase possibly supporting the Majorana modes to a trivial phase accompanied by CdGM-type excitations. Doping the bulk TI can, for instance, induce such a change of the μ. The energy gap related to the CdGM excitations prevents their mixing with the topological Majorana states at the ends into the vortex line threading through the bulk SC, and consequently reduces the QP poisoning. The magnitude of the $\mu_c = \hbar v_F \sqrt{\frac{m}{B}}$, where v_F, m, and B are the Fermi velocity, effective mass, and the curvature of the bands at the nontrivial to trivial band crossing, respectively [146]. The determination of the μ_c would be sensitive to a band topology characteristic and may be understood through a related topological invariant such as the Z_2 index [17, 157]. Such a description also lends itself to the idea that a tuning of the μ_c perhaps, even in the absence of band inversion typical to a TI, could yield Majorana modes in a nontopological SC. Here, sufficient S–O interaction energy (greater than that of the SC gap) is necessary along with nonequivalent points in the band structure of the related *trivial* superconductor.

3.5.1 *Probing the Possibility of Majorana Modes in the Vortices of Se-Based SCs*

An early example of possible Majorana states in the vortices of SC has been related to a prototypical 3-D TI [41], that is, Bi_2Se_3, with SC induced through proximity to an *s*-wave superconductor [153], that is, $NbSe_2$. A quintuple layer (QL) of Bi_2Se_3, constituted from two Bi layers encompassed by three Te layers was the base unit. It was indicated that the topological surface states were formed only after 3 quintuple layers (3QL) were deposited on the SC. However, at 3QL it was found that the high-bulk carrier density *submerged* the unconventional characteristics from the topological surface states. However, from 4 QL to 6 QL, the topological surface state behavior was more visible. A corresponding ZBP was observed near the vortex center with a related peak splitting with increasing distance from the center. It was indicated that the local DOS measured by the STM, as would be related to the QP states, would be maximal at (/away from) the vortex core in conventional (/topological) SCs and result in a peak splitting. Consequently, the Majorana mode would correspond to a true zero-energy state with minimal-related DOS. The influence of vortex interaction was probed where at fields greater than 0.1 T, a reduction of the ZBP intensity was observed. However, with the given STM energy resolution of ~0.2–0.3 meV, the full-width half-maximum (FWHM) of the ZBP peak seemed quite large.

The resolution of 0.2 meV essentially sets an energy scale to probing the "zero-energy" aspect of the Majorana mode. Moreover, energies related to CdGM excitations must also be distinguished, which further mandates that at the temperature (T) of operation, would correspond to an energy: $k_B T << \varnothing_{SC}^2 / E_F$. Consequently, a large $\varnothing_{SC}^2 / E_F$ ratio guarantees clear distinction of the nontopological from the topological states in the SC gap. While a large Δ_{SC} is indicative of high T_c superconductors, a large E_F is typical from the intrinsically large carrier concentrations which is true in the case of most SCs. Consequently, the $\varnothing_{SC}^2 / E_F$ ratio is typically small and of the order of 0.001. A much larger value of the ratio of ~0.7 meV, was indicated in the *intrinsic* topological superconductor: $FeTe_{0.55}Se_{0.45}$ (with a T_c of ~14.5 K, Δ_{SC} ~ 1.8 meV, and E_F ~ 4.4 meV) and was touted as a motivation for the possibility of identifying Majorana modes at the ends of the related vortex core lines [158].

In this context, the iron-based SC, for example, α-FeSe, has been indicated to be stabilized in a *well-defined* SC phase when the materials were prepared with intentional Se deficiency [159]. Consequently, the favorable addition of substitutional Te (for Se): Fig. 3.14a, in $FeTe_{0.55}Se_{0.45}$, has the advantage of increasing the T_c by ~7 K. Additionally, the lowering of the energy of the p_z band (from the chalcogenide atom) induces interactions with the d_{xz}/d_{yz} bands from the Fe, and causes a topological band inversion along the Γ-Z direction in the corresponding Brillouin zone: Fig. 3.14b.

The S–O coupling, from the heavy atoms (Se, Te), further results in a related band gap that could host Dirac-cone like surface states, of helical texture, exhibiting an SC gap (of ~1.8 meV), of *s*-wave character, as was revealed through high-resolution laser ARPES [67]—(angle-resolved photoemission spectroscopy): Fig. 3.14c. With respect to the detailed energy distribution in the bulk SC, the band structure with a related SC gap at the k_F, cf., Fig. 2.1c, is formed in the bulk: Fig. 3.14d. Subsequent inter-band scattering [67] in the bulk was hypothesized to induce, through proximity, *s*-wave SC into a topological surface state (TSS)—yielding *intrinsic* topological SC. The TSS are spin-resolved due to their helical nature, where spin is locked to the momentum, that is, the two opposite momenta or traveling directions have two different spin orientations. Such spin resolution by transport is not feasible in the bulk SC due to its topologically trivial character, and the absence of spin–momentum locking: Fig. 3.14d.

The Δ_{SC} and E_F values were obtained through fitting and correlating the experimental ZBP intensity to a theoretical spatial profile (from solving the BdG equations) expected for the Majorana state. Consequently, the obtained zero-energy topological state could be clearly discriminated from the expected CdGM states at a vortex core separated by ~0.7 meV as well as the thermal broadening at ~0.17 meV (= 3.5 k_BT, with a T ~ 0.55 K). In STM probing of a vortex, created under an applied magnetic field of 0.5 T: Fig. 3.15a, a pronounced ZBP was observed in the vortex center with a SC background–related spectrum indicating a Δ_{SC} ~ 1.8 meV. The presumed tunneling of electrons from the STM tip into the Majorana mode has been associated with the ZBP. Proceeding from one side of the vortex to the other side, the development of the ZBP in the vortex region is observed in the spectral variation of the differential conductance in Fig. 3.15b and more in detail in Fig. 3.15c. It was indicated that the width of the Majorana mode (as indicated through the full width at half maximum: FWHM) was limited by the instrumentation resolution at ~0.3 meV and is comparable with the "thermal smearing"—as indicated in Fig. 3.15d. The associated spatial variation of the Majorana mode–related intensity profile is shown in Fig. 3.15e. However, it was claimed that the FWHM could be increased depending on the background of the SC, for example, an increased impurity background could yield a *softer* SC gap and reduced protection of the putative Majorana modes from the ambient as well as related QP poisoning. The ZBP was

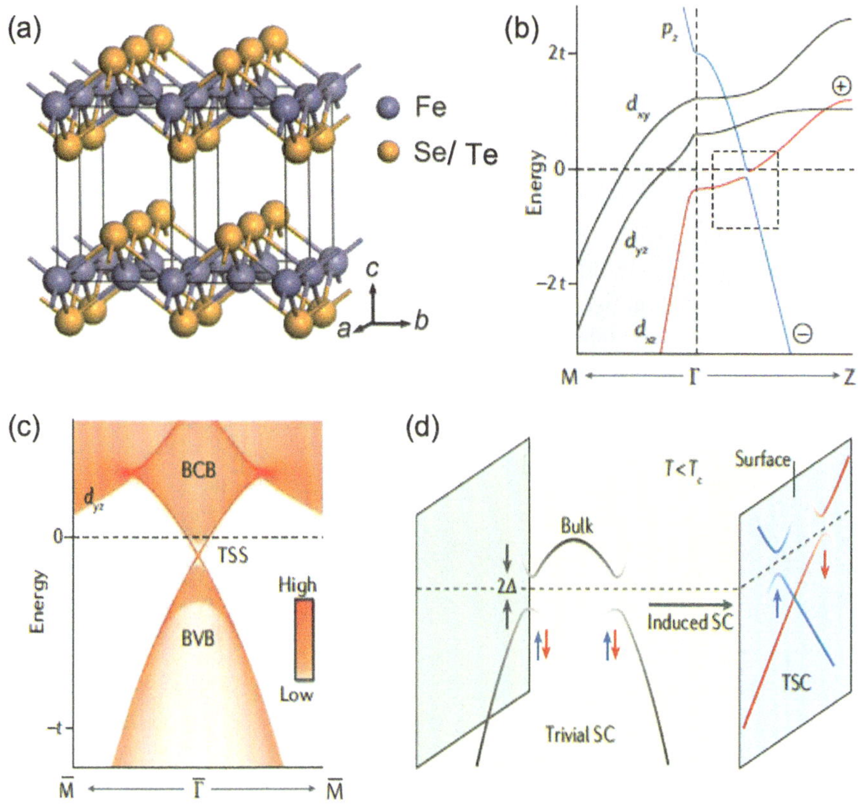

Fig. 3.14 *A basis SC for investigating vortex states* (**a**) The crystal structure of FeSe—from [159], and the (**b**) related band structure—from [67]. The substitution of Te for Se lowers the energy of the p_z bands whose interaction with the d_{xz}/d_{yz} bands yields a band inversion related to a topological phase transition. The + and – denote the parity of the bands to spatial inversion, (**c**) from [67]. The band structure depicted in the surface Brillouin zone (BZ), showing the topological surface states (TSS) of helical (with spin–momentum locking) character hosted in the S–O-coupling-induced bandgap between the bulk valence band (BVB) and the bulk conduction band (BCB), (**d**) from [67] – at temperature (T) < T_c, the SC transition temperature, the bands in the bulk SC (depicted on the *left*) show the expected SC bandgap (of magnitude = $2\Delta_{SC}$) at $\pm k_F$. The projection of the bulk bands onto the surface (depicted on the *right*), coupled with induced SC into the helical channels, resolves the spin polarization (depicted by the blue and red arrows). Majorana modes can then be associated with the corresponding zero-energy states in the surface gap

found to be robust over the range of 0.1 T–6 T (yielding magnetic lengths in the range of ~2500 nm to 10 nm [75]) as would be expected from the spatial length scale of ≥ 10 nm, indicated in Fig. 3.15e.

While the observation of a *true* Majorana mode is still suspect due to the unclear identification of the influence of the background as well as possible magnetic impurities, a rather interesting feature of such work is the possible higher temperature of operation. The ZBP intensity was found to vanish at temperatures close to ~3 K, less allied to the energy scales related to the SC gap and, perhaps, more related to the background. The related reduction of the amplitude of the ZBP is expected, as the energy gap between the Majorana and CdGM states of the $FeTe_{0.55}Se_{0.45}$ is smaller than the SC gap and hence the possibility of thermally excited quasiparticles *poisoning* the Majorana states would be plausible. Consequently, issues related to both disorder with respect to the resolution coupled with other physical effects must be considered.

Various other sources of the ZBP were sought to be eliminated broadly through magnetic field tuning [158], that is, based on (i) Andreev reflection processes inducing phase conjugation, (ii) influence of impurities which may also modify QP interactions yielding a conductance maxima predicated on Kondo effect–related phenomena and/or the softening of the SC gap due to sub-gap conductance, (iii)

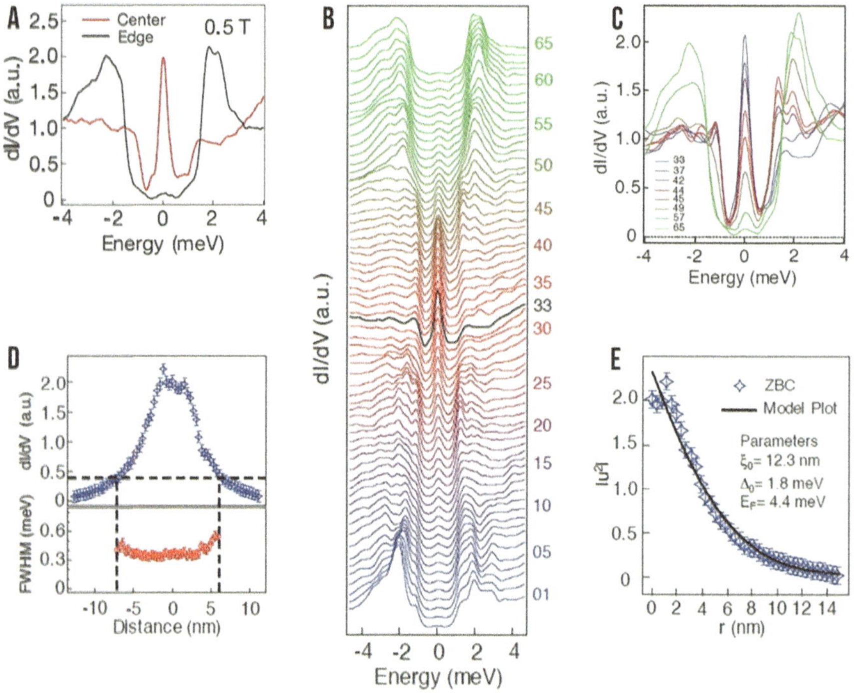

Fig. 3.15 *Zero-bias peaks (ZBP) at the ends of vortex cores.* (Adapted from [158]): (**a**) A sharp ZBP observed in differential conductance measurements at the center of the SC vortex (in red), whereas features related to the bulk SC, with an associated gap of ~1.8 meV are seen in the spectrum at the edge of the vortex (in blue) (**b**) The spectral variation of the differential conductance from the *left* (bottom curves) to the *right* (top curves) along the vortex. The black curve, in the center, represents the center of the vortex. (**c**) The conductance variation of representative curves from (**b**) indicate the robustness of the ZBP. (**d**) Peak height extracted from (**b**) as a function of the distance from center of vortex (in blue) along with the full-width at half maximum (FWHM) of the peaks at each location (in red). (**e**) Comparison of the spatial decay of the ZBP height, from the vortex center, with theoretical models

Josephson supercurrents which may occur at zero applied bias, (iv) weak anti-localization-related effects again with respect to constructive/destructive interferences, etc., all of which were thought to be ruled out.

As indicated previously, the conductance of the Majorana mode should be immune to the strength of the tunnel coupling, from an external electrode, and should yield G_o. This is in accord with previous observations in NWs using external electrodes (Sect. 3.1.2 as related to Fig. 3.1) where the quantization of the ZBP was shown from tuning the tunnel coupling. The evidence of a Majorana mode through STM on vortices in SCs was sought through variation of the tip–sample spacing, which would be the relevant parameter for the tuning of the tunnel coupling and parameterized through G_N (= I_t/V_s, where I_t was the tunneling current and V_s was the setpoint voltage). Optimizing the overlap of states between the tip and the sample could also result in less smearing of the conductance peaks. The vacuum barrier inherent in experiments interfacing the STM tip and the sample would be relevant, along with the tunneling of electrons from the tip to specific zero-energy states/possible Majorana modes. The advantage of tunneling through vacuum would be a more rapid variation of the potential profile compared to a smoother change and greater variability that would be prevalent through defect mediated tunneling in the solid state. QP poisoning from the ambient could be also reduced. It is possible that Andreev reflection would be involved in the tip–sample space, corresponding to the SC/N/SC configuration, cf. Fig. 2.2, modulating the tunneling characteristics.

3.5.2 *Obtaining a Mode with a Zero-Bias Peak (ZBP) Conductance ~* G_o

In associated low-temperature STM studies on $FeTe_{0.55}Se_{0.45}$: Fig. 3.16a, such a *nearly quantized conductance plateau* [76] was claimed to have been observed in the ZBP-related intensity. A double barrier wherein the electrons and holes interfere destructively and realize *perfect* transmission, irrespective of the strength of the tunnel barriers, would be the basis of such quantization of the conductance to G_o. As the strength of the coupling is enhanced, a saturation of the conductance of the ZBP at the core to ~0.65 G_o and in some cases even approaching G_o was observed; in contrast the conductance increases monotonically outside the SC gap: Fig. 3.16c. The saturation was achieved when the G_N itself was greater than ~0.3G_o and for increasingly larger values of the coupling. Moreover, other features ascribed to the CdGM states were also observed to monotonically increase with the G_N, in contrast to the behavior of the ZBP. Here, the breaking the electron–hole symmetry in the suspected CdGM state was implicated in the absence of a clear plateau in the measured conductance. However, examination of different vortices seemed to yield widely varying values of the saturation conductance, even as low as ~0.1 G_o and seems to be correlated with an increasing FWHM of the ZBP. Quasiparticle poisoning was implicated in the broadening and the somewhat related reduction of the saturation/plateau conductance.

It was claimed that the convolution of the instrumental broadening of ~0.12 meV (at 78 mK) had the influence of lowering the obtained conductance from close to G_o to ~$0.65G_o$. Interestingly, lock-in amplifier modulation voltage was also implicated in enhancing the instrumental broadening and reducing the amplitude of the Majorana conductance peak! The reported work was interesting in that it presents the ambiguity of clearly identifying the ZBP—possibly hosting the Majorana mode. At the very outset, the aspect of fitting to somewhat arbitrary parameters (as in the lock-in voltage) to obtain the desired mode, as well as the value of G_o, is unsatisfactory. Moreover, the peaks were fit assuming a Voigt profile—corresponding to a probability distribution itself formed by the convolution of a Gaussian distribution to the Lorentz distribution, and seemingly related to the instrumental broadening as well as the *intrinsic* Majorana signal, respectively. Such partitioning does not possess clear justification. However, the extreme fragility of the mode may also be apparent, in the necessity of such fits.

In the above experiments several constraints are seen under which even the possibility of the Majorana mode must carefully be delineated and indicate criteria that are not clear in scope. First, the thermal broadening should be suppressed through very low temperatures of experimentation. The low temperatures (sub-Kelvin levels) also allow the SC gap to grow close to its maximum value (as defined at 0 K). The Majorana state at zero energy must clearly be separated from other quasiparticle excitations) in the vortex by an energy of $\varnothing^2_{SC}/E_F$. Associated with the favorable large SC gap would be a short ξ, which implies a localized environment for the resident modes, and which may be useful to reduce vortex pinning effects. For instance,

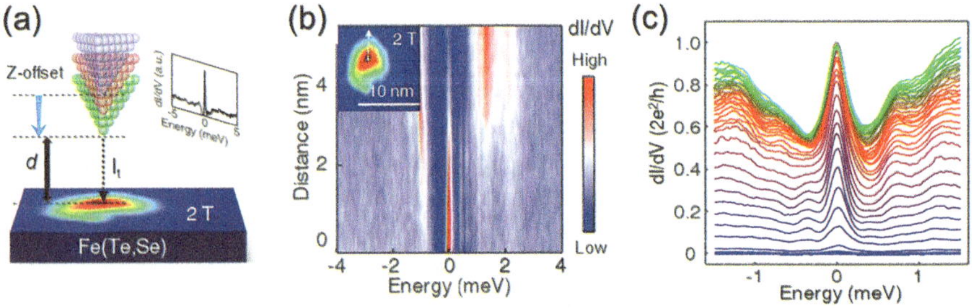

Fig. 3.16 (**a**) Schematic arrangement of STM probing of the vortex of a $FeTe_{0.55}Se_{0.45}$ SC with variable height (Z-offset–based tunnel coupling of the STM to the vortex core. The current (I_t) is a monitor of the coupling at a given set-point voltage (V_s). The differential conductance (*G; dI/dV*) may be measured at different tip–sample distances (*d*). A conductance map of the vortex (obtained with a perpendicular applied magnetic field of 2 T at 78 mK) is shown on the sample surface. (**b**) The (*dI/dV*) map along various distances from the core (located at 0 nm), along the dashed white arrow in the *inset*. (**c**) A sampling of the variation of the ZBP at the center of several vortices indicates a *G nearly* approaching the conductance quantum: G_o in a few cases. Instrumental broadening as well as quasiparticle poisoning were indicated as plausibly responsible for the deviation of the *G* from G_o. (Adapted from [76])

the ξ may be in the order of ~3 nm in the FeSe-based topological superconductors, in comparison to the value of ~30 nm in $Bi_2Te_3/NbSe_2$-based materials. It would be very pertinent then to investigate higher T_c SCs for both increased temperature of operations as well as to understand whether there would be a correlation with the increased localization of the Majorana mode.

3.5.3 The Influence of Defects in the Vortex Cores

In similar vein, another report of a *robust* Majorana mode was indicated [160] through vortex core interrogation in an alternate FeSe-based SC, that is, $(Li_{1-x}Fe_x)OHFeSe$, $x \sim 0.84$, with a high T_c (~42 K), and estimated to be in a topologically nontrivial phase. While the specific deployment of this compound was based on obtaining stoichiometry, and hence reducing disorder, the E_F was reported to be quite large at ~57 meV. With a corresponding Δ_{SC} of the order of 7 meV–15 meV, the $\varnothing_{SC}^2/E_F$ would be estimated to be ~2 meV. Such a range in the Δ_{SC} arises from the substantial gap anisotropy, that is, deviation from pure *s*-wave SC, with a pair of coherence peaks—effectively yielding two SC gaps ($\Delta_{SC,1}$ and $\Delta_{SC,2}$)—tentatively ascribed to the topological surface band and the bulk band (related to the M point of the Brillouin zone in the SC), respectively: Fig. 3.17a. With an $x \sim 0.25$, an SOC-induced gap was indicated, through combined density functional theory (DFT) and dynamical mean field theory (DMFT)–based computations and corresponding Angle-Resolved Photo-Emission Spectroscopy (ARPES) experiments, to yield a topologically nontrivial phase hosting gapless topological states (which may correspond to the Majorana modes) on the surface.

The application of a magnetic field (of up to 10 T) revealed vortices on the surfaces of two types, that is, one set pinned at defects (with dark spots at the center, with a small differential conductance (*dI/dV*) probed by the STM) and another set corresponding to a free vortex (with a large *dI/dV* at the center). The former set was influenced by impurity sites associated with Fe dimer-like substitutional defects. STM observations yielded *dI/dV* as a function of distance from the center of a vortex core: Fig. 3.17c. Here, an inner set of approximately five closely spaced low energy peaks was observed close to the vortex center along with another set of spectrally broad states, presumably allied to the CdGM manifold, away from the center. With an STM resolution of ~0.36 meV, the inner set of five peaks were fit to multiple Gaussian peaks, and an associated full width half maximum (FWHM) of ~0.6–0.8 meV for the ZBP was obtained: Fig. 3.17d. Through a fitting of the energy separation between the two outer (/inner) peaks surrounding the ZBP to Δ_2 and Δ_1, as in Fig. 3.17a, their origination from the topological surface state (/*trivial* bulk band) was proposed. Given that the CdGM states are supposed to be spaced ~2 meV, the ZBP was then indicated to represent a topologically non-trivial *zero-energy* state. Moreover, the CdGM states have a tendency to move toward/accumulate at

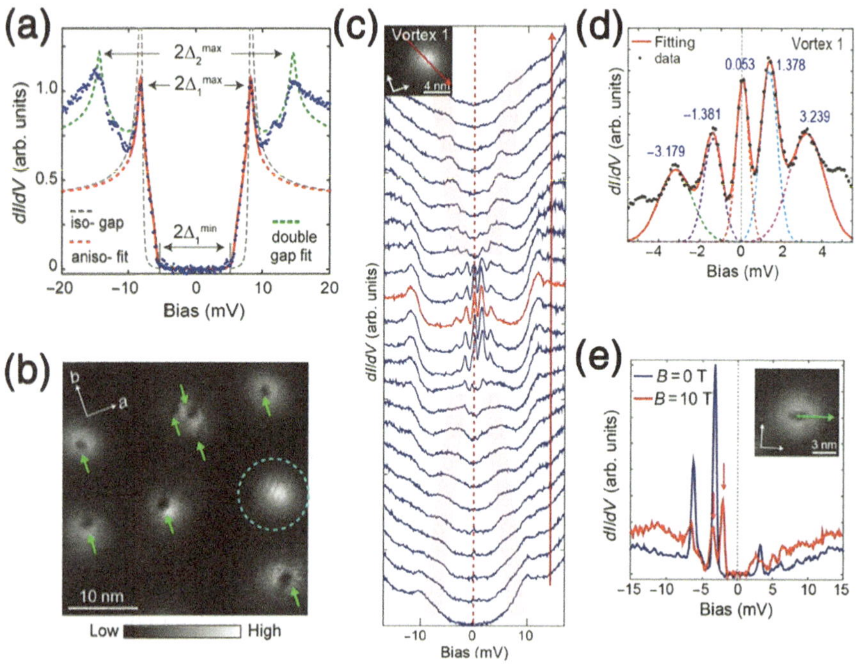

Fig. 3.17 *Finding Majorana modes in vortex cores.* (Adapted from [160]) (**a**) Spectral variation of the differential conductance map of the ($Li_{0.84}Fe_{0.16}$) OHFeSe measured through STM, indicating the possibility of two SC gaps, that is, an anisotropic Δ_1 related to the gap of a topological surface state and another gap Δ_2 related to the bulk SC gap. The ***black***, *red*, and *green* dotted lines represent fits assuming an isotropic, anisotropic, and the possibility of two anisotropic SC gaps, respectively. (**b**) A zero-bias conductance map of an area of the SC surface, with an applied B_{ext} of 10 T, indicating pinned (*green arrows*)—typically with dark spots at the center, and free vortices (blue dashed circle). (**c**) The differential conductance mapping across a vortex (from the *left* to the *right* in the *inset* vortex image, with corresponding spectra from *bottom* to *top* in the plots), as a function of tip bias. The presence of multiple peaks at the center and pairs of broader peaks away from the center (shaded in pink) were observed. The broad bands far away from the vortex core are ascribed to Caroli–de Gennes–Matricon (CdGM) features. (**d**) A closer resolution of the vortex core states. Two of the inner states, surrounding the ZBP, at ~ ± 1.38 mV are ascribed to a trivial bulk bands while the ZBP itself and the two outer peaks (at ~ ± 3.2 mV) are indicated to arise from the topological surface states. The red curve represent multiple Gaussian peak fits (*dotted lines*) to the data (black dots). The peak bias values of each Gaussian fit are indicated. (**e**) The differential conductance mapping as a function of bias at the center of a pinned vortex (*inset*) at two different magnetic fields shows no ZBP, but instead yield impurity-related peaks at finite bias, presumably from magnetic Fe dimer defects [Use another figure 5(b), from [112] instead!]

the gap edge away from the vortex center, as also indicated by the shaded pink area in Fig. 3.17c:

While both free and impurity-pinned vortices were observed in differential conductance mapping, only free vortices (defect-free and unpinned) were repeatedly seen to contain the robust Majorana zero modes, related to the ZBPs. The conductance mapping on the pinned vortices also yields peaks close to zero-bias, as in

Fig. 3.17e but are more typically of a double-peaked feature as related to the split impurity nature of the associated Fe dimer defects. While there is an obvious emphasis on producing clean vortex lattices for a clearer delineation of the Majorana modes, such spectral signatures offer an obvious scheme from confounding states. However, deconvolution of such impurity peaks from the CdGM-related spectral features must also be considered.

While stoichiometry was initially supposed, the disparate atoms/ions in the compound still seem to yield a considerable number of defects/impurity states, related to the Fe, which pin the vortices. Moreover, the specific effect of the gap anisotropy on the energy broadening (of both the ZBP as well as nearby energy states) was unclear and raises questions on the expected FWHM of the ZBP. The possibility of constructive/destructive interferences as well as the influence of Fe proximate to the core was not well considered, especially considering that ZBPs are also evident at Fe-related defect sites. Moreover, the speculated coupling between the CdGM modes with the impurity modes generally leads to confusion in the interpretation of the STM spectra. A related study on Fe (Te,Se) indicated the role of interstitial iron impurities (IFIs) as the possible location of the spatial features, where the ZBP may be found[162]. Such a zero-energy state was found to be robust to applied B_{ext} as large as 8 T, and assigned to a MBS. However, the observed STM conductance was substantially smaller than the G_o. Here again, the interaction of magnetic impurities with bulk SC states, such as the Shiba or the YSR states (Sect. 2.5) must be considered.

More detailed STM studies on Fe adatoms on top of a $FeTe_{0.55}Se_{0.45}$ surface, Fig. 3.18a, did indicate the possibility of spin-independent Coulomb scattering and the related formation of YSR in-gap states [162]. The low values of the G of the order of $10^{-4}G_o$ observed in the tunneling was used to rule out Andreev scattering and possible ABS. Here, of a total of 46 studied adatoms, spectroscopy indicated YSR states on ~19 adatoms, with a conductance minimum, corresponding to say Fig. 3.18b, while a ZBP was observed on ~14 adatoms, with a conductance maximum, corresponding to say Fig. 3.18c. The YSR states have an intrinsic particle–hole symmetry, with respect to the energy, as also indicated in Fig. 2.4b, and are related to the aspect of the scattering of the Cooper pairs by the Fe atoms. The difference in the amplitude of the YSR states on either side of the zero-bias may be due to the asymmetry of the spin-dependent Coulomb scattering. A decay length of ~0.5 nm for the YSR states, modeled as magnetic dipoles, was posited [162] and related to the SC coherence length (~2 nm, here) as well as the relative position of the YSR states within the SC gap (Δ_{SC} ~ 1.8 meV). Spin-polarized tunneling into the states was demonstrated, with a pronounced difference in the G obtained through changing the direction of the magnetic field on the STM tip. However, a clear spin signature, with respect to the spin polarization, was absent with respect to the Fe atoms showing the ZBP, as in Fig. 3.18c, and indicated further in Fig. 3.19, where there is a clear lack of spin polarization in the tunneling spectroscopy.

The absence of a response to spin polarization in the ZBP would then preclude an associated MBS. It was then indicated that a ZBP could result from YSR states closely spaced in energy and the mixing of the related quasiparticles of opposite

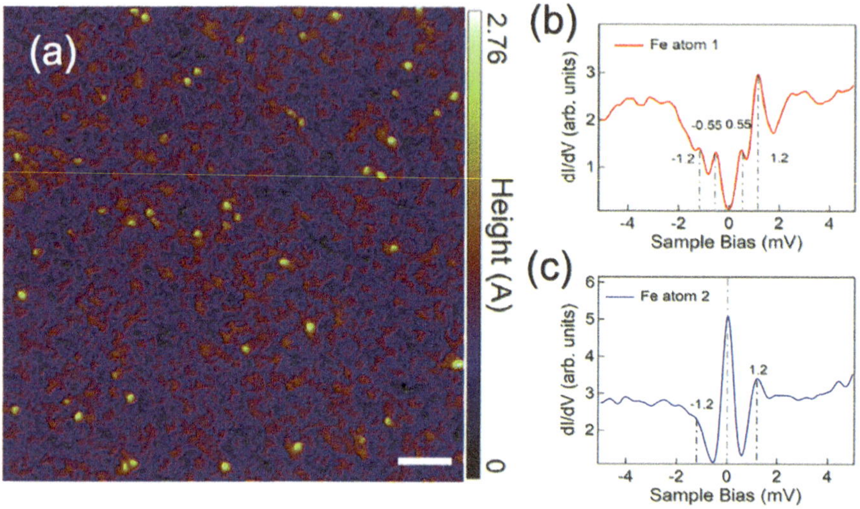

Fig. 3.18 (**a**) Topography of the surface of a cleaved $FeTe_{0.55}Se_{0.45}$ surface with Fe atoms (*yellow*), with a *scale bar* of 5 nm. Spectroscopy conducted on the top of two distinct Fe adatom, indicating either a (**b**) minimum, *or* a (**c**) maximum in the conductance. In (**b**), the YSR states are indicated at ± 1.2 mV. (From [162])

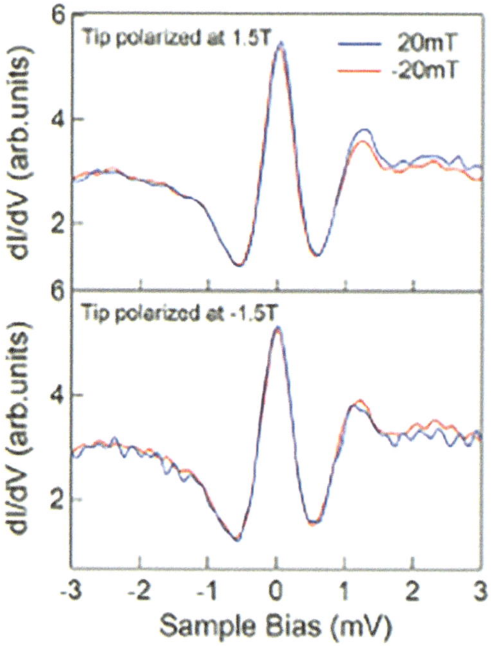

Fig. 3.19 *Lack of spin polarization of the ZBP observed on an Fe adatom through tunneling spectroscopy*. Magnetic field measurements using an STM tip magnetized to ±1.5 T does not indicate an underlying spin sensitivity for the adatom. A small bias field of ±20 mT was applied to orient the magnetic moment of the adatom

spin character, close to the E_F, and not due to the Majorana states. Such an *accidental* coincidence of the states at zero bias would then be another consideration to be ruled out for a definitive establishment of the existence of the Majorana modes.

In summary, STM-related spectroscopy of the vortex cores in intrinsically topological SCs may not yet have provided clear evidence of the presence of the Majorana modes. There is yet substantial advantage in the STM-based methodologies, given the (a) absence of coupling between two diverse materials, (b) presence of vacuum as a *clean* tunneling barrier, and (c) the promise of higher temperature operation, as high T_c may be corelated with topological SCs.

Chapter 4
Issues Related to Determination of Majorana Fermion Related Modes

It was initially expected that a clear signature of a Majorana mode could be obtained through a conductance quantization value of G_o at zero bias, in a suitably prepared topological SC. However, such an idealized experimental observation has generally not been observed and remains one of the biggest obstacles to the identification of the MBS. In addition to weak coupling and limitations from experimental resolution, for example, finite temperature (leading to peak broadening) and finite system size (precluding a clear Majorana delocalization), only "approximate quantization", [82, 163] has been generally obtained. Even then, the MBS at the two ends of the one-dimensional system such as a NW has been found to be nonequivalent [163].

Figure 4.1 indicates a recent experiment where such a difference between the conductances at the left- and right-hand sides of the NW, that is, $G_L \neq G_R$, is apparent. The measurement configuration with is indicated in Fig. 4.1a and incorporates an InSb NW (with left and right contacts) and into which SC has been induced from an overlaid NbTiN, along with connecting leads. Two tunnel gates, at the left and right, are used to create tunnel barriers from the contacts to the topological SC, and serve to inject electronic current and create quasiparticle states in the SC. Another gate (termed the S-gate) was used for controlling the electron density and consequently to control the overlap the MZM modes on the two ends. The conductance peak, corresponding to the *left* side were observed at (i) a large magnetic field (B_{ext}) of ~1 T, (ii) are persistent with respect to the B_{ext}, and (iii) close to a magnitude of G_o—all indicative of Majorana-like characteristics. The peak was also independent of the tunnel barrier coupling with the experiments conducted in a regime of large tunnel coupling—with respect to the thermal energy. The necessity for the B_{ext} to be oriented as exactly parallel to the NW was noted. However, under the same conditions where the *left* conductance peak was observed: Fig. 4.1b, there was no corresponding ZBP on the *right* side: Fig. 4.1c. While the length of the NW is rather small, at ~400 nm, such experiments still indicate the necessity of three-wire measurements for the clear demarcation of the MBS at the ends. Moreover, the observed ZBP (only on the *left* side) could not now be definitively ascribed to a MBS/

© The Author(s), under exclusive license to Springer Nature Switzerland AG 2022
P. Bandaru, S. Natani, *Topological States for New Modes of Information Storage and Transfer*, https://doi.org/10.1007/978-3-030-93340-1_4

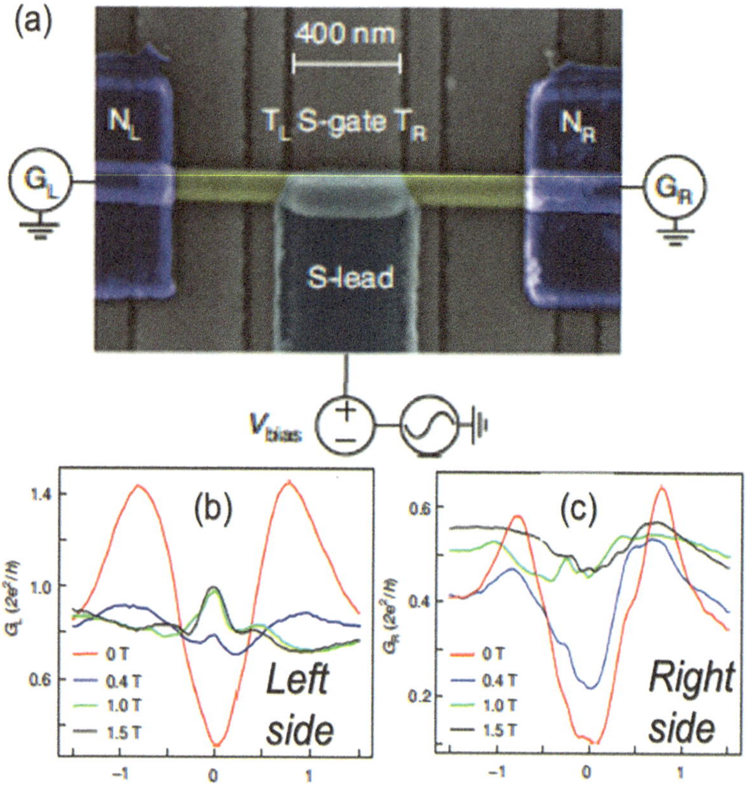

Fig. 4.1 From [163] Investigation of the Majorana zero modes at the left side and right side of a NW. (**a**) An InSb NW into which SC has been induced from an overlaid NbTiN—connected through the S-lead. The NW contacts on the two sides are of normal metal (Pd), and indicated as N_L and N_R. The tunnel gates (T_L and T_R) create tunnel barriers from the contacts to the topological SC, and serve to inject electronic current and create quasiparticle states in the SC. The S-gate is for controlling the electron density related to the S-lead and may be used to control the overlap the MZM modes on the two ends. The conductances of the two sides of the device (G_L and G_R) are measured simultaneously, using two lock-in amplifiers, and recorded in (**b**) and (**c**), respectively. It is apparent that the G_o has not been achieved and that $G_L \neq G_R$. Moreover, the same experimental conditions which are used to produce a ZBP on the *left* side do not seem to yield a ZBP on the *right* side

Majorana mode but may also be associated with trivial ABS as well as disorder-induced energy levels. Further, the observation of multiple Majorana modes, with identical signatures in multiple regions of the systems, would provide further proof.

In light of the above discussion, an unambiguous identification of a Majorana mode has not yet been achieved. Considering the previous discussion, the ZBP by itself may not be a clear discriminator of the modes. Consider the obtained resolution of the ZBP to within the thermal limit of ~2 μeV @ 20 mK, which has been achieved. The splitting (/non-splitting) of the ZBP has then been associated with non-MBS (/MBS) states. Most of the other possibilities have been sought to be eliminated, through their dependence on the B_{ext}. For example, there is a possibility of the conflating of the Kondo effect (which yields a resistance maxima) with the peaks observed at $B_{ext} \neq 0$. This may occur trivially due to a rescaling/error in the measured B_{ext}. The methodology related to such scaling is not generally obvious. It has also been indicated that the MBS may be considered a subset of the total set of zero-bias related states, that is, in the ABS state manifold [51]. Consequently, significant attention and discrimination would be necessary for MBS identification. Indeed, a major issue that needs further investigation are the contributions, from the mid-gap states, as previously discussed in Sect. 2.5. A few relates issues precluding clear identification will be discussed next.

4.1 Sub-Gap States with Finite Zero-Bias Conductance, Confused with Majorana Modes

It was indicated that MBS may not be clearly distinguished in any NW of *finite* length, due to the formation of ps-ABS states [52], and consequently alternate *nonlocal* measurements may need to be performed. While there is some merit to such a recommendation, given the delocalized character of the Majorana states, it is unclear yet as to what measurements would provide a definitive proof as opposed to "smoking gun"–type signatures. From the point of view of transport measurements, three-terminal measurements for the identification of *both* the individual Majorana state [163], and subsequent correlation of the results were recommended [52]—see Fig. 4.1. Here, assuming equivalent transport, say through the tunnel barriers formed at the contact—SC proximitized NW, or at the QD–NW interface, an MBS would be revealed with a conductance of G_o. The ABS states may also shift in response to an external magnetic field, crossing the E_F and through a quantum phase transition evolve to spin polarized states. Such a crossing was seen to produce zero-bias peaks (ZBPs) in the conductance that may again be confused with Majorana modes [164]. In the presence of magnetic impurities, it has already been indicated, for example, as in Sect. 3.5.3 or Fig. 3.17, that closely spaced YSR states may also yield large ZBP. There is also considerable uncertainty as related to the role of unintentional disorder, for example, as related to magnetic impurity atoms and related appearance of in-gap states of the SC, and related materials systems. A major issue is whether a sufficiently localized wavefunction at the impurity would yield signatures expected from an MBS. While significant efforts have been expended to rule out the related possibilities, for example, as indicated in the resolution of the Shiba/YSR states [101, 116] indicated in Sect. 2.5, further insight into the role of impurities on the ABS manifold would be helpful.

The recent invoking of the possibility of charge-related Coulomb sub-gap states in modulating the conductance at zero-bias must also be considered [165]: Sect. 2.5. Here, quantum dot–SC island electrostatic interactions were indicated to produce zero-bias crossings of the resultant hybridized states. Such Coulomb states may be *distinguished* from other electron–hole symmetric Andreev states or the YSR states, experimentally through a related breaking of the symmetry with respect to the applied positive and negative bias. From an applied magnetic field point of view, the related spin triplet states are distinguished from the nominal singlet states expected in the typical SC-related electron coupling.

4.2 The Influence of Disorder on Conclusively Determining the Majorana States

A significant effort has been expended in reducing both intrinsic and extrinsic disorder in experimental systems, from the aspect of the establishment of (a) Majorana localization length as well as (b) reducing in-gap states in a SC for obtaining a *hard* gap. Indeed, a topological phase may only be defined if the disorder is sufficiently small. The scaling would be considered with respect to the ratio of a relevant length scale, for example, length (L) of the NW and the SC coherence length (ξ). It has been indicated, for instance, that with any finite L NW, the Majorana states combine into fermionic excitations of mean energy E_o that is exponentially small as related to L/ξ and with a log-normal probability distribution [166]. Such an aspect is interesting for positing that the Majorana mode distribution is due to the accumulation of several accumulating contributors, as is typified through the log-normal attribute [167], as well as "large sample to sample fluctuations" [166]. Moreover, it is expected that disorder, that is, through nontopological or trivial defects, such as nonmagnetic impurity atoms or vacancies, would increase the effective Majorana localization length (ζ), implying a spread-out character for the related modes. Indeed, a relationship $\frac{1}{\zeta}=\frac{1}{\zeta_o}-\frac{1}{2l}$ has been posited for p-wave paired SCs, where ζ_o is the effective localization length in the *clean* limit and l being an impurity induced mean free path. A *critical disorder strength* may then be defined at the limit $2l = \xi$; with the implication that when $2l > (/<)\ \xi$; that the topological (/nontopological) phase prevails. Coupled with the increase in the ζ, such considerations indicate that *minimization of the disorder in a very long wire could possibly provide adequately protected Majorana modes*.

Another encouraging aspect, provided the disorder is below a threshold, is that weak disorder could even reduce the ζ [168]. The foundational argument is based on modifying the phase change occurring on scattering through considering in more detail the type of disorder, for example, magnetic vs. nonmagnetic. For the latter kind, isotropic scattering implies the conventional argument, where increased disorder does enhance the ζ. Here, the electron spin polarization is preserved and the Δ_{SC} could be ascribed, as in the conventional BCS sense, to s-wave-like character. However, with magnetic-impurity scattering, the extent of spin polarization would be reduced (due to scattering between spin polarized channels), effectively diminishing the spin-character of the topological phase. One may then attribute the SC gap in the latter case to be p-wave like ($= \Delta_{SC,p}$). Such a deeper delving into the nature of the topological phase (with a combination of Δ_{SC} and $\Delta_{SC,p}$, in various proportions) is appealing from enabling a better understanding of topological SCs, for example, with analogy to the hypothesized $s - + d$- wave character in high T_c SCs [109, 169]. The interplay of Δ_{SC} and $\Delta_{SC,p}$ would then determine the extent of topological SC and the localization of the resultant Majorana modes, for example, if $\Delta_{SC,p} < \Delta_{SC}$ the BCS coherence length $\xi_{SC} > \xi_{SC,p}$, with isotropic scattering where the ζ proportionally increases with disorder, while with if $\Delta_{SC,p} > \Delta_{SC}$, the $\xi_{SC} < \xi_{SC,p}$, with ζ varying non-monotonically with the disorder. In the latter case, there is a minimum in the ξ as a function of the disorder, with the implication that disorder may be beneficial in decreasing the l as well as the ζ. It would be important in future studies to *distinguish disorder* through the ratio of the magnetic and nonmagnetic/potential contributions [168]. It was found, for instance, that as this ratio increases, for a given disorder, that the decrease in the ξ diminishes, with a decreased susceptibility to the disorder and an enhanced stability of the topological phase. Consequently, the possibility of a clear delineation of the MBS would be enhanced.

4.3 The Influence of a Small Device Size in Precluding Proper Majorana Mode Localization and Identification

At the very outset, the Majorana modes correspond to fully localized modes and would ideally be situated at the ends of an infinite-sized one-dimensional system, such as a very long NW. However, from a practical perspective, such a long NW is not desirable, in that it would allow for more disorder and also runs counter to the aspects related to miniaturization. With respect to the latter, present-day transistor electronics have devices in the sub 5 nanometer range [170], and it would not bode well for a future technology to be predicated on lengths that are orders of magnitude larger! However, while shorter NWs are desirable, for example, there is a possible reduction of the number of Majorana conflating sub-gap states due to a diminished tunneling length [171], they would for instance preclude the exponential protection of the Majorana modes that may be present in 1-micron-long NWs. Moreover, the aspect of simultaneously determining the Majorana modes at the ends of the wires, for greater confidence in the observations, would be reduced in shorter wires.

Chapter 5
Suggestions for Future Experiments

Major issues preventing the realization of widespread quantum information science (QIS) and quantum computation (QC) in the solid state are the stability and the fidelity of the quantum bit (*qubit*) used for encoding information. Decoherence due to the interaction of the quantum state with the environment can destroy the quantum information. The aspect of identifying and harnessing Majorana modes directly addresses this issue by probing the feasibility of topological states (TS) which are inherently protected from local perturbations, and will preserve the robustness of quantum information storage in a related *qubit*. We deem that the following set of aims and criteria would be of high importance for realizing such aims.

5.1 Identification of Suitable Superconductors for Hosting Unique TS

It would be much desirable to resolve the existing ambiguity related to TS and probe unique TS amenable to storing quantum information. The goals should be to revisit the determination as to whether specific SC types, for example, those exhibiting *p*-wave SC-related features, are necessary for TS. This should be done through theoretical investigations of (a) the core states in the vortices of SCs; (b) the relevance of the electron spin and intrinsic magnetization in inducing the TS; in addition to (c) a reexamination of extant TS-related observations, considering the deviation from the effects from true SC states. While (a) and (b) provide a basis for interpreting initial experimental inquiries, (c) would help in aligning the priorities of the quantum information science (QIS) community. A search for new materials exploiting recently described high-throughput machine-learning approaches [173] would also be exciting, in addition to the consideration of the possibility of using high transition temperature (T_c) superconductors (HTS), with $T_c \geq 4$ K, where modern notions of topological SC may be applicable such as $Cu_xBi_2Se_3$ [172] or the Fe

© The Author(s), under exclusive license to Springer Nature Switzerland AG 2022

P. Bandaru, S. Natani, *Topological States for New Modes of Information Storage and Transfer*, https://doi.org/10.1007/978-3-030-93340-1_5

chalcogenides [174] ($T_c > 40$ K). The possibilities [41] of obtaining unique energy states in other HTS such as $YBa_2Cu_3O_{7-\delta}$ (YBCO) or $Sr_2CaCu_2Bi_2O_8$ (Bi2212) should also be of high priority.

5.2 Probing the Topological States Through Alternative STM Modalities

From the viewpoint of establishing the very existence of MBS, it would be highly desirable to directly probe the TS rather than infer their presence and influence through somewhat indirect and invasive electrical measurements. The use of Josephson Scanning Tunneling Microscopy (JSTM) [175] to identify the noninteracting energy states, for example, as may be found in the magnetic vortex core could directly prove the existence of zero-energy TS. The tunneling of electrons, from a known and understood set of states—through a well-defined vacuum potential—to a possible TS would definitively identify the character of the latter state. Both quasiparticle tunneling as well as Josephson tunneling could be deployed. Utilizing a superconducting tip on the STM will increase the energy resolution of this spectroscopy by *at least two orders of magnitude* by using the sharp gap edge [100]. It is noted that the resolution of the JSTM spectroscopy would not be limited by thermal effects and in principle, could be the resolution of the measuring voltmeter.

In experiments on the vortex cores, a normal/non-SC tip has been mostly used for the STM spectroscopy, where an enhanced *DOS*, associated with a substantial ZBP was observed, Fig. 3.13c. However, the ZBP appears quite broad with a width of ~2 meV (note that the measurement temperature of 1.8 K corresponds to ~0.15 meV). Utilizing a SC tip on the STM will substantially increase the energy resolution of the spectroscopy by *orders of magnitude* through using the sharp SC gap edge. The related high-energy resolution measurements, not subject to thermal broadening limits, will help better differentiate the various energy modes and allow access to possible states inside a vortex core, such as the CdGMs, other core-associated ABS and geometric resonance–associated states, and finally "*zero in*" on the possible TS levels. The measurements with an STM with the SC tip would deploy the Josephson effect (a result of the SC electron pair wavefunction coherence across a barrier between two superconductors). Considering only Josephson tunneling between two *s*-wave SCs, the pair coupling is well understood as the product of the junction resistance R and the critical current I_c, and is a fixed quantity related to the SC gap. However, Josephson tunneling between a conventional SC, for example, that exhibiting *s*-symmetry and an unconventional SC, for example, that exhibiting *d*- or *p*-symmetry or a topological SC, is problematic and violates fundamental quantum mechanics principles related to the orthogonality and exclusiveness of the wave functions. Consequently, high-energy resolution JSTM coupled with wavefunction parity specificity would be an incisive probe into the nature of the TS, for example, whether the TS is present, if it is of a particular symmetry, and if electron spin is or is not important. The Josephson tunneling technique is hence particularly suited to probing the TS state in a pristine SC and is more sensitive to indirect inference on the TS through electrical contact-based measurements in proximity effect–induced SC. Future work with focus on a careful examination and energy spectroscopy of the of the vortex core of *s*-wave and unconventional SCs with respect to its capability for hosting TS would be of high import.

Such experiments should be closely tied to computational modeling efforts through careful study of the local electronic structure around the vortex cores of the commonly considered *p*-wave SC (e.g., $Cu_xBi_2Se_3$) and assign a likelihood for the possibility of true TS. The deployment of effective low-energy models and topological analyses [176] accounting for the multi-band, multi-valley bulk and surface SC characteristics could be used to gain insight into the nature of the topological SC and its sensitivity to the presence of disorder. Additional large-scale lattice simulations [155] to examine the nature of the bound states in vortices and the interaction between states localized in two different vortices are desired. The local electronic structure should be computed for the vortex cores and the associated QP states and correlated with JSTM observations. Further modeling to analyze the energy spectra and the symmetries of the zero-mode wavefunctions will help determine if they satisfy the property of Majorana modes. Additionally, the influence of dimensionality could be probed through consideration of the effect of layer thickness of $FeTe_{1-x}Se_x$ and nanowires [177] or 1-D $NbSe_3$.

5.3 Implementation of Electrode-Based Schemes for Modulation and Readout of the Topological States

It would be highly desirable to integrate and implement electrode-based schemes, amenable to scale up, for the manipulation of the zero-energy states, once they are identified. The true identification of a TS immune to environmental disturbance, and useful for quantum information processing can only be definitively indicated through the implementation of specific braiding Schemes [29, 30, 36, 178]. For instance, vortex cores may be induced through an applied magnetic field and their subsequent interaction could be probed through specific thin film architectures and electrical voltage–based measurements. The control of the vortex motion could be done through inducing potential gradients using materials modification. Here, ion beams with sub-nanometer resolution could be used to create tracks/guiding grooves for the motion of the vortices.

After specific identification of unique states amenable to hold and manipulate quantum information (as indicated in Sect. 1.3) efforts should be focused manipulating the states in a scalable electrode-based device platform. First, control over a given area (S), the number of vortices $(n) = B_{ext}S/\Phi_0$ using an applied magnetic field (B_{ext}) would need to be obtained. The *number* as well as the relative distances of the vortices could be tuned, through varying the B_{ext} [147]. The vortex itself could be moved and relocated by the application of small currents—through guiding vortex motion along electromagnetic potential gradients [179]. Given that a vortex may be modeled as a magnetic flux tube (carrying a flux of Φ_o), the application of a current would induce a transverse force on the vortex and cause its motion [180]. The moving vortex would concurrently induce an electrical field and related voltage transverse to the motion, which can be measured to monitor individual or relative motion of the vortices. Practically, the guiding of vortices could be done through potential profiles induced through patterned grooves over a range of length scales.

The vortex guiding tracks may be fabricated by irradiating the SC materials with a finely focused beam of helium ions with a diameter of ~0.35 nm. It has been shown that the obtained electrical properties are stable for years [181] with demonstrated fabrication of *in-plane* narrow quantum tunnel barriers through such patterning [40]. Generally, in materials sensitive to disorder such as complex chalcogenides and related high temperature SCs, ion irradiation can be used to dramatically change the electrical characteristics, for example, Fig. 5.1a shows resistivity as a function of temperature for irradiated YBCO thin films [182], while the fabrication of *in-plane* narrow quantum tunnel barriers through such patterning is a schematic of the resolution of the helium ion beam. It could be ensured, for example, through JSTM studies that the guiding groove would only provide a track for the vortex motion and not trap the vortex.

The subsequent motion and interaction of vortices along the guiding grooves could be facilitated through imposed electrical currents. The voltages that would be measured consequently will yield insights into the nature of the quantum information. Pulse-driven vortex motion along the grooves could also be used to probe evolution of braiding.

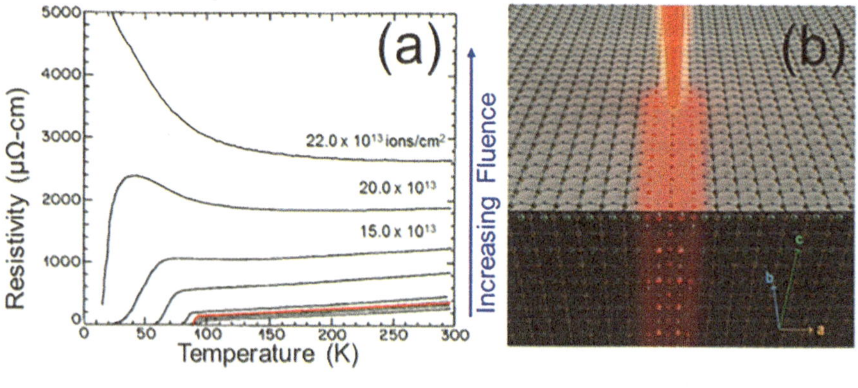

Fig. 5.1 **(a)** The electrical resistivity of superconductors, for example, $YBa_2Cu_3O_{7-\delta}$ (YBCO) thin films, may be tuned through ion irradiation (dose indicated in ions/cm^2 units), through **(b)** *sub-nanometer direct write* He ion beam irradiation [182]. Such a technique could be used to form the electrical potential profiles/grooves, to guide vortex motion, and hence move the possible TS or Majorana modes for subsequent qubit manipulation

5.4 Braiding Schemes Would Demonstrate the Utility of TS for Quantum Information

The existence as well as the utility of a TS may be verified through particular braiding operations and could be accomplished through the mutual interaction of the vortices. As previously discussed, for example, in Sect. 3.5.1, the core of a SC vortex yielding the ZBP could contain the TS. Consequently, moving the vortices around each other would be equivalent to a braiding operation and could be investigated. For this purpose, a multilayer SC based arrangement, Fig. 5.2, with two SCs sandwiching a thin (~1 nm) insulating/normal barrier, could be deployed. The difference of the interaction between the TS would be manifested in a relative phase (ϕ) and yield a voltage (V) in accord with the Josephson relation: $d\phi/dt = 2\,eV/\hbar$. An s-wave SC, say $NbSe_2$, could constitute the *bottom* layer of the heterostructure. A guiding groove could be fabricated on the layer through helium ion beam–induced patterning on the SC—as outlined earlier in Fig. 5.1b. An insulator (e.g., AlO_x) could be subsequently deposited by atomic layer deposition (ALD). Another SC (with s–/p–/d- characteristics) could be deposited (through thin film sputtering) or exfoliated onto the insulator layer. One may choose, for this layer, an unconventional SC, such as (1) one with distinct topological characteristics, conferred through large spin–orbit interaction, say amorphous Bi, (2) *p-wave* type SC such as Sr_2RuO_4, or (3) $Cu_xBi_2Se_3$ or the Fe chalcogenides. Helium ion beam patterning could be used to pattern a smaller guiding groove on the *top* SC, of a size (~50 nm) required to confine a single vortex.

Electrical currents applied to the bottom SC would move the vortex flux line around the trapped vortex and could be used to demonstrate vortex core state/quantum flux line braiding. The current guided vortex movement could be done through either applied (a) electrical fields, or (b) polarized electromagnetic waves. The polarization of the related electrical fields could be used to move the vortices either clockwise or counterclockwise. Concomitant with vortex motion, the zero-energy/ZBP-related modes, say, as observed in the $NbSe_2$ could be guided *around* the bound vortex in the top SC. The *twisting* of the flux lines may be correlated to a unitary operation and related non-Abelian exchanges, forming the crux of quantum information exchange [2].

A single braiding of two modes, for example, implements a Hadamard transform, and repeating it gives the identity operation. The robustness of the braiding operations and exchange could be seamlessly and nonlocally monitored through recording the voltages (AC and DC). Generally, the interaction between the vortex motion in the *top* and the *bottom* superconductors corresponds to magnetic coupling and a transformer-like action. This would yield, for example, a voltage which yields a measure of the extent of *classical/non-quantum* interaction. The exchange of *quantum information* would be indicated through an *additional* voltage enabled through the relative phase differences pertinent to quantum interactions.

The voltages should be measured in two ways for consistency and compatibility, that is, (i) between the top and bottom SCs, as well as (ii) on the top SC hosting the

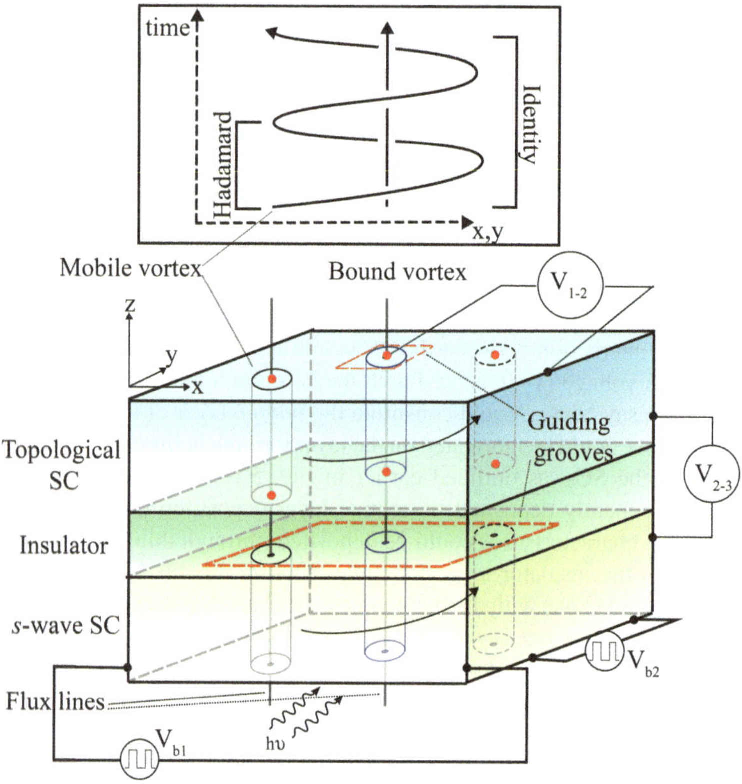

Fig. 5.2 Schematic of a plausible experimental design, for probing topological states associated with interacting zero-bias peak (ZBP)-related features and energy states in SC vortexes. A topological SC (*e.g.*, amorphous Bi/Sr_2RuO_4) is separated by an insulator (*e.g.*, AlO_x) from an *s*-wave SC (*e.g.*, $NbSe_2$). An applied magnetic field (*B*) in a patterned area tunes the number and spacing of the vortexes. Single vortices may be generated at a given point [148]. The flux line representing the vortex line would be continuous through the two SCs and the insulator. The vortex motion proceeds along the *guiding grooves*—fabricated through He-ion beam patterning, *cf.* Fig. 2(b), and is induced through currents provided through bias voltages (V_{b1} and V_{b2}) or *polarized* electromagnetic waves (as indicated at the bottom). The movement of the bottom vortex around the top vortex would be a braiding operation if the vortex harbors a TS. The possible topological states are indicated as red dots. The phase difference due to the vortex motion would be transduced to a *measured* electrical voltage between the top and bottom SCs ($V_{2\text{-}3}$) or on the top SC itself ($V_{1\text{-}2}$), using an electrode placed on the vortex [183] and inform on the nature of the energy mode interactions. (*Top inset:* A single braiding of two modes of the vortex, for example, implements a Hadamard transform, and repeating it gives the identity operation)

TS. In (i) the measured voltage ($V_{2\text{-}3}$) would be the sum of the voltages due to (a) the conventional SC–I–SC coupling (of magnitude ~1.5 Δ_{SC}/e[101], (b) that generated through moving one magnetic flux tube around another flux tube, as well as from (c) moving defined energy states, such as the ABS or the CdGM states around each other, and (d) the TS around each other. The contribution from (d) would be related to the TS. We previously considered the difference of the $V_{meas(i)}$ and (a) to be related to the deviation of the SC from *s*-wave characteristics [100] and a measure of the order parameter [169]. As indicated previously, zero angular momentum (and related zero energy states) may always be present in trajectories where the order parameter is real [63], and the rotation of the vortices may yield voltage signals that would be able to distinguish energy levels with or without angular momentum.

An alternate measured voltage (V_{1-2}) on the top layer of the SC would be between an electrode on the bound vortex and the bulk SC. Such a measurement protocol has been previously proposed for the testing of non-Abelian statistics [183]. An extensive correlation and corroboration of the electrical voltages using various test structures with an aim to providing a *measure* of the ratio of classical *to* quantum information exchange is imperative.

Subsequent to a proper understanding of vortex motion guiding and energy state interaction from such simple arrangements, various spatial topologies [118] should be used to vary and measure the correlated phase profiles and accompanying voltages, paving the way to deduce higher-order interactions, relevant to braiding operations in anyonic states [30]. One must also consider the possibilities related to measurement-induced braiding operations and related QC [184], for example, through interferometric measurements related to vortex motion [185]. Such proposed schema should be capable of extension, so as to enable qubit operations. For instance, it should be investigated if a nonlocal encoding of the energy states, as part of the ZBP, would indeed be the *key* to ensure adequate protection of sets of states [30] to local perturbations. Further work on communication between the qubits, based on encoding and interfacing the phase degree of freedom, onto, say, surface acoustic wave–based devices [186, 187] would be of high interest for the engineers. In summary, vortex core–based topological states (TS) can possibly serve as the basis for a new variety of decoherence-free qubits and subsequent operations involving the *braiding* of the TS states for performing a Hadamard operation involving quantum information exchange.

Chapter 6
Outlook

The aspect of identifying new modalities of energy states, and their subsequent manipulation for hitherto inconceivable technologies, lies at the core of the experimental search for the Majorana modes. Indeed, a subsequent set of milestones based on Majorana-based quantum computation has been recently indicated [188], and involves mainly the following: (i) the identification for the interaction of the Majorana modes, through charge-based sensing; (ii) validation of the related *qubit*; and (iii) the demonstration of unique non-Abelian statistics, through the qubit interactions. However, the modes must first be clearly identified and delineated. We have indicated that many experiments on a wide variety of materials systems have been tantalizingly close to such identification. However, the continuing efforts in the field point to the lack of certainty. Furthermore, spatial localization of the Majorana modes have also not been probed directly in many material systems. Indeed, the following *five* criteria were recently indicated as *key* to the identification of the Majorana modes, that is,

(i) Zero-bias peaks (ZBP) which persist over a range of magnetic fields (B_{ext}). An increasing B_{ext} causes a bandgap closure and subsequent band inversion, as indicated in Fig. 2.1. Consequently, the ZBP occurs at a specific B_{ext}. The related voltage range/energy width of the ZBP (= $e\delta V$) would need to be restricted and smaller compared to the increase in the Zeeman energy for a unit increase δB_{ext}, that is, $g\mu\delta B_{ext} > e\delta V$.

(ii) Ensure that the B_{ext} is perpendicular to the intrinsic B_{SO} for maximal ZBP amplitude. This may be verified through angle-dependent measurements, where the amplitude changes cyclically from a minimum (/maximum) when the B_{ext} is parallel (/perpendicular) to B_{SO}.

(iii) The ZBP is evident over a range of electrical and magnetic fields, that is, enable a phase diagram for comparison with theory.

(iv) The ZBP splits into two peaks after the two Majorana modes at the ends annihilate each other. This could be shown through measuring oscillations of the

© The Author(s), under exclusive license to Springer Nature Switzerland AG 2022

P. Bandaru, S. Natani, *Topological States for New Modes of Information Storage and Transfer*, https://doi.org/10.1007/978-3-030-93340-1_6

number of peaks by varying many parameters (such as the electrical or magnetic fields) and showing that there are indeed two peaks after the Majorana-related particles annihilate each other.

(v) Finally, the Majorana modes, posited to exist in pairs, should be simultaneously observed, for example, at the ends of a NW. The overlap of the modes may be regulated through the application of bias voltage. Such an aspect is particularly relevant for braiding operations and relevant to indicating anyonic statistics. Indeed, this aspect may be a *smoking gun* for the proof of the existence of the modes (S. Frolov).

In this context, STM-based investigations which interrogate particular states would be of high priority. The complexity of isolating Majorana modes and gathering spatial information of the modes should be the biggest challenges in the near future. After confirming the existence of the MZMs, the next step would be to develop a working MZM-based qubit system and investigate its possible utility for quantum information science. This would require the control and subsequent interaction/braiding [30, 178] of at least *four MZMs* and implementation of simple quantum logic gate, for example, the Hadamard gate [2]. The demonstration of the non-Abelian anyonic statistics [30] would be a most exciting prospect for the utility of the MZMs. Subsequently, the full knowledge related to the implementation of practical devices, for example, qubit redundancy, error control and correction, noise reduction, etc., may be brough to bear on the utilization of the MZMs. Presently, quantum error correction codes use many qubits to correct the error related to a single qubit [2], and topological qubits incorporating the Majorana modes may be useful, as they have a higher degree of intrinsic suppression of decoherence, implying a reduced emphasis on error correction. As it has been indicated that universal quantum computation may not be feasible with the MZMs [29], integration with alternate/nontopological qubit systems may be necessary and would need to be probed as well.

While much of the focus on the discovery of the MZMs has been using TSCs, other connections related to the occurrence of MZMs could be investigated. For instance, spin triplet supercurrents through Josephson junctions with a ferromagnet in between two SCs [189] could manifest odd spin-triplet SC-induced in the ferromagnetic material [190]. Indeed, induced SC in FM chains has been previously considered (Sect. 3.3). Finally, while many suggestions have been made related to the applications of MZMs for logic and memory, the possibilities related to quantum sensing remains an open area for breakthroughs.

Another major issue plaguing the field of Majorana discovery has been a "*reproducibility crisis*" [191], where it was indicated that many key experiments, performed by over one hundred groups internationally, have not been confirmed. While detection with high fidelity is difficult, it seems imperative that complete data representation without bias and subject to scrutiny from the community be done. Indeed, the "*key discoveries have yet to be made*" [191]. We may then expect that a clear identification of the Majorana mode would be akin to the discovery of the electron in the latter part of the nineteenth century and bring forth the next century of new undreamt-of technologies.

References

1. B.Q. Lv et al., Experimental discovery of Weyl Semimetal TaAs. Phys. Rev. X **5**(3), 031013 (2015). https://doi.org/10.1103/PhysRevX.5.031013
2. M.A. Nielsen, I.L. Chuang, *Quantum Computation and Quantum Information* (Cambridge University Press, Cambridge, 2011)
3. R. P. Feynman, *Simulating Physics with Computers* (1982)
4. S.A. Aaronson, *Quantum Computing Since Democritus* (Cambridge University Press, Cambridge, 2013)
5. R.D. Mattuck, *A Guide to Feynman Diagrams in the Many-Body Problem*, 2nd edn. (Dover Publications Inc., Mineola, NY, 1976)
6. D.J. Griffiths, *Introduction to Quantum Mechanics*, 2nd edn. (Cambridge University Press, Cambridge, 2017)
7. R. Hoffmann, *Solids and Surfaces, a chemist's View of Bonding in Extended Structures* (VCH Publishers, New York, 1988)
8. C. Kittel, *Introduction to Solid State Physics* (Wiley, New York, 1996)
9. W. Shockley, On the surface states associated with a periodic potential. Phys. Rev. **56**(4), 317–323 (1939). https://doi.org/10.1103/PhysRev.56.317
10. D. Long, *Energy Bands in Semiconductors* (Interscience, New York, 1968)
11. R.E. Prange, S. Girvin, *The Quantum Hall Effect* (Springer, New York, 1987)
12. Spin-orbit coupling effects in zinc blende structures. Phys. Rev. **100**, 580 (1955)
13. E.I. Rashba, Spin-orbit coupling goes global. J. Phys. Condens. Matter **28**, 421004 (2016)
14. M.S. Dresselhaus, G. Dresselhaus, A. Jorio, *Group Theory: Application to the Physics of Condensed Matter* (Springer, Berlin/Heidelberg, 2008)
15. K. He, Y. Wang, Q.-K. Xue, Topological materials: Quantum anomalous hall system. Annu. Rev. Condens. Matter Phys. **9**(1), 329–344 (2018). https://doi.org/10.1146/annurev-conmatphys-033117-054144
16. Y. Ando, L. Fu, Topological crystalline insulators and topological superconductors: From concepts to materials. Annu. Rev. Condens. Matter Phys. **6**(1), 361–381 (2015). https://doi.org/10.1146/annurev-conmatphys-031214-014501
17. L. Fu, C.L. Kane, Topological insulators with inversion symmetry. Phys. Rev. B – Condens. Matter Mater. Phys. **76**(4), 045302 (2007). https://doi.org/10.1103/PhysRevB.76.045302
18. A.B. Bernevig, T.L. Hughes, *Topological Insulators and Topological Superconductors* (Princeton University Press, Princeton, 2013)
19. D. Vanderbilt, *Berry Phases in Electronic Structure Theory: Electric Polarization, Orbital Magnetization and Topological Insulators* (Cambridge University Press, Cambridge, 2018)

© The Author(s), under exclusive license to Springer Nature Switzerland AG 2022

P. Bandaru, S. Natani, *Topological States for New Modes of Information Storage and Transfer*, https://doi.org/10.1007/978-3-030-93340-1

20. M. Tinkham, *Introduction to Superconductivity*, 2nd edn. (Dover Publications Inc., Mineola, 2004)
21. C. Kittel, *Introduction to Solid State Physics*, 7th edn. (Wiley, New York, 2003)
22. S. Blundell, T. Lancaster, *Quantum Field Theory for the Gifted Amateur* (Oxford University Press (OUP), Oxford, 2014)
23. F. Wilczek, Majorana returns. Nat. Phys. **5**, 614 (2009)
24. M. Fox, *Optical Properties of Solids* (Oxford University Press, New York, 2001)
25. J. Bardeen, L.N. Cooper, J.R. Schrieffer, Theory of superconductivity. Phys. Rev. **108**(5), 1175–1204 (1957). https://doi.org/10.1103/PhysRev.108.1175
26. E. Majorana, *A Symmetric Theory of Electrons and Positrons(*)* (1937)
27. E. Merzbacher, *Quantum Mechanics*, 2nd edn. (Wiley, New York, 1970)
28. L.I. Schiff, *Quantum Mechanics* (McGraw Hill, New York, 1968)
29. M. Leijnse, K. Flensberg, Introduction to topological superconductivity and Majorana fermions. Semicond. Sci. Technol. **27**(12), 124003 (2012). https://doi.org/10.1088/0268-1242/27/12/124003
30. C. Nayak, S.H. Simon, A. Stern, M. Freedman, S. Das Sarma, Non-abelian anyons and topological quantum computation. Rev. Mod. Phys. **80**, 1083 (2008)
31. D.P. Divincenzo, *Topics in Quantum Computers* (1996)
32. T. Van Duzer, *Principles of Superconductive Devices and Circuits*, 2nd edn. (Pearson/Addison Wesley, New York, 1998)
33. R. Eisberg, R. Resnick, *Quantum Physics of Atoms, Molecules, Solids, Nuclei, and Particles*, 2nd edn. (Wiley, New York, 1985)
34. A. Kitaev, *Unpaired Majorana fermions in quantum wires* (2000), pp. 1–16. https://doi.org/10.1070/1063-7869/44/10S/S29
35. L. Pauling, *The Nature of the Chemical Bond* (Cornell University Press, Ithaca, 1959)
36. J. Alicea, New directions in the pursuit of Majorana fermions in solid state systems. Reports Prog. Phys. **75**, 076501 (2012)
37. A.P. Mackenzie, Y. Maeno, p-wave superconductivity. Phys. B Condens. Matter **280**(1–4), 148 (2000)
38. E.F. Talantsev et al., p-wave superconductivity in iron-based superconductors. Sci. Reports **9**(1), 1–13 (2019). https://doi.org/10.1038/s41598-019-50687-y
39. L. Capriotti, D. J. Scalapino, and R. D. Sedgewick, "Wave-vector power spectrum of the local tunneling density of states: Ripples in a *d*-wave sea," Phys. Rev. B, vol. 68, no. 1, p. 014508, Jul. 2003, doi: https://doi.org/10.1103/PhysRevB.68.014508.
40. S.A. Cybart et al., Nano Josephson superconducting tunnel junctions in YBa2Cu3O(7-δ) directly patterned with a focused helium ion beam. Nat. Nanotechnol. **10**, 598 (2015)
41. L. Fu, C.L. Kane, Superconducting proximity effect and Majorana Fermions at the surface of a topological insulator. Phys. Rev. Lett. **100**, 096407 (2008)
42. X.-L. Qi, S.-C. Zhang, Topological insulators and superconductors. Rev. Mod. Phys. **83**, 1057 (2011)
43. S. Yonezawa, *Bulk Topological Superconductors* (2016). https://arxiv.org/pdf/1604.07930.pdf.
44. B.D. Cullity, C.D. Graham, *Introduction to Magnetic Materials* (Wiley-IEEE Press, New York, 2008)
45. R.P. Feynman, R.B. Leighton, M. Sands, *The Feynman Lectures in Physics* (Addsion Wesley, New York, 1964)
46. J.J. Sakurai, J. Napolitano, *Modern Quantum Mechanics*, 2nd edn. (Addison-Wesley, San Francisco, 2011)
47. H. Kroemer, C. Kittel, *Thermal Physics* (W.H. Freeman and Co., New York, 1980)
48. C.H.L. Quay et al., Observation of a one-dimensional spin–orbit gap in a quantum wire. Nat. Phys. **6**(5), 336–339 (2010). https://doi.org/10.1038/nphys1626
49. C. Beenakker, Serach for Majorana fermions in superconductors. Annu. Rev. Condens. Matter **4**, 113 (2013)

50. V. Mourik, K. Zuo, S.M. Frolov, S.R. Plissard, E.P.A.M. Bakkers, L.P. Kouwenhoven, Signatures of Majorana fermions in hybrid superconductor-semiconductor nanowire devices. Science **336**(6084), 1003–1007 (2012). https://doi.org/10.1126/science.1222360
51. M.T. Deng et al., Majorana bound state in a coupled quantum-dot hybrid-nanowire system. Science (80-.) **1557**(December), 1–17 (2016). https://doi.org/10.1126/science.aaf3961
52. C. Moore, T.D. Stanescu, S. Tewari, Two-terminal charge tunneling: Disentangling Majorana zero modes from partially separated Andreev bound states in semiconductor-superconductor heterostructures. Phys. Rev. B **97**(16), 165302 (2018). https://doi.org/10.1103/PhysRevB.97.165302
53. C. Moore, C. Zeng, T.D. Stanescu, S. Tewari, Quantized zero-bias conductance plateau in semiconductor-superconductor heterostructures without topological Majorana zero modes. Phys. Rev. B **98**(15), 155314 (2018). https://doi.org/10.1103/PhysRevB.98.155314
54. S.M. Sze, K.K. Ng, *Physics of Semiconductor Devices* (Wiley, Hoboken, 2006)
55. S.M. Sze, *Semiconductor Devices: Physics and Technology*, 2nd edn. (Wiley, Singapore, 2003)
56. A.V. Balatsky, I. Vekhter, J.X. Zhu, Impurity-induced states in conventional and unconventional superconductors. Rev. Mod. Phys. **78**(2), 373–433 (2006). https://doi.org/10.1103/RevModPhys.78.373
57. F. von Oppen, Y. Peng, F. Pientka, *Topological Superconducting Phases in One Dimension* (Oxford University Press (OUP), Oxford)
58. H. Shiba, Classical spins in superconductors. Prog. Theor. Phys. **40**(3), 435–451 (1968). https://doi.org/10.1143/ptp.40.435
59. A. Jellinggaard, K. Grove-Rasmussen, M.H. Madsen, J. Nygård, Tuning Yu-Shiba-Rusinov states in a quantum dot. Phys. Rev. B **94**(6), 064520 (2016). https://doi.org/10.1103/PhysRevB.94.064520
60. J. Carlos, et al., *Coulombic Subgap States* (2021)
61. W.J. Tomasch, Geometrical resonance and boundary effects in Tunneling from superconducting in. Phys. Rev. Lett. **16**(1), 16–19 (1966). https://doi.org/10.1103/PhysRevLett.16.16
62. J.M. Rowell, W.L. McMillan, Electron interference in a normal metal induced by superconducting contracts. Phys. Rev. Lett. **16**(11), 453–456 (1966). https://doi.org/10.1103/PhysRevLett.16.453
63. D. Rainer, J.A. Sauls, D. Waxman, Current carried by bound states of a superconducting vortex. Phys. Rev. B **54**(14), 10094–10106 (1996). https://doi.org/10.1103/PhysRevB.54.10094
64. L. Fu, C.L. Kane, Josephson current and noise at a superconductor/quantum-spin-hall-insulator/superconductor junction. Phys. Rev. B **79**(16), 161408 (2009). https://doi.org/10.1103/PhysRevB.79.161408
65. B. Seradjeh, E. Grosfeld, Unpaired Majorana fermions in a layered topological superconductor. Phys. Rev. B **83**(17), –174521 (2011). https://doi.org/10.1103/PhysRevB.83.174521
66. Y.S. Hor et al., Superconductivity in CuxBi2Se3 and its implications for pairing in the undoped topological insulator. Phys. Rev. Lett. **104**(5), 057001 (2010). https://doi.org/10.1103/PhysRevLett.104.057001
67. P. Zhang et al., Observation of topological superconductivity on the surface of an iron-based superconductor. Science **360**(6385), 182–186 (2018). https://doi.org/10.1126/science.aan4596
68. J.A. Sauls, Andreev bound states and their signatures. Philos. Trans. R. Soc. A Math. Phys. Eng. Sci. (2125), 376, 20180140 (2018). https://doi.org/10.1098/rsta.2018.0140
69. C.-X. Liu, J.D. Sau, T.D. Stanescu, S. Das Sarma, Andreev bound states versus Majorana bound states in quantum dot-nanowire-superconductor hybrid structures: Trivial versus topological zero-bias conductance peaks. Phys. Rev. B **96**(7), 075161 (2017). https://doi.org/10.1103/PhysRevB.96.075161
70. A.Y. Kitaev, Unpaired Majorana fermions in quantum wires. Phys. Uspekhi **44**(10S), 131–136 (2001). https://doi.org/10.1070/1063-7869/44/10S/S29
71. J.D. Sau, S. Tewari, R.M. Lutchyn, T.D. Stanescu, S. Das Sarma, Non-abelian quantum order in spin-orbit-coupled semiconductors: Search for topological Majorana particles in solid-state systems. Phys. Rev. B **82**, 214509 (2010)

72. E.J. Schwalbach, P.W. Voorhees, Phase equlibrium and nucleation in VLS grown nanowires. Nanoletters **8**, 3739–3745 (2008)
73. J.D. Sau, R.M. Lutchyn, S. Tewari, S. Das Sarma, Generic new platform for topological quantum computation using semiconductor heterostructures. Phys. Rev. Lett. **104**(4) (2010). https://doi.org/10.1103/PhysRevLett.104.040502
74. R.S. Muller, T.I. Kamins, M. Chan, *Device Electronics for Integrated Circuits*, 3rd edn. (Wiley, New York, 2003)
75. S. Datta, *Electronic Transport in Mesoscopic Systems* (Cambridge University Press, New York, 1995)
76. S. Zhu et al., Nearly quantized conductance plateau of vortex zero mode in an iron-based superconductor. Science (80-.). **367**(6474), 189–192 (2020). https://doi.org/10.1126/science.aax0274
77. S. Datta, *Quantum Transport: Atom to Transistor* (Cambridge University Press, New York, 2005)
78. M.T. Deng, C.L. Yu, G.Y. Huang, M. Larsson, P. Caroff, H.Q. Xu, Anomalous zero-bias conductance peak in a Nb-InSb nanowire-Nb hybrid device. Nano Lett. **12**(12), 6414–6419 (2012). https://doi.org/10.1021/nl303758w
79. A. Das, Y. Ronen, Y. Most, Y. Oreg, M. Heiblum, H. Shtrikman, Zero-bias peaks and splitting in an Al-InAs nanowire topological superconductor as a signature of Majorana fermions. Nat. Phys. **8**(12), 887–895 (2012). https://doi.org/10.1038/nphys2479
80. P. Krogstrup et al., Epitaxy of semiconductor-superconductor nanowires. Nat. Mater. **14**(4), 400–406 (2015). https://doi.org/10.1038/nmat4176
81. H. Zhang et al., Quantized Majorana conductance. Nature **556**(7699), 74–79 (2018). https://doi.org/10.1038/nature26142
82. A.C. Potter, P.A. Lee, Multichannel generalization of Kitaev's Majorana end states and a practical route to realize them in thin films. Phys. Rev. Lett. **105**(22), 227003 (2010). https://doi.org/10.1103/PhysRevLett.105.227003
83. T.D. Stanescu, R.M. Lutchyn, S. Das Sarma, Majorana fermions in semiconductor nanowires. Phys. Rev. B – Condens. Matter Mater. Phys. **84**(14), 144522 (2011). https://doi.org/10.1103/PhysRevB.84.144522
84. R.M. Lutchyn, T.D. Stanescu, S. Das Sarma, Search for Majorana fermions in multiband semiconducting nanowires. Phys. Rev. Lett. **106**(12), 127001 (2011). https://doi.org/10.1103/PhysRevLett.106.127001
85. J. Liu, A.C. Potter, K.T. Law, P.A. Lee, Zero-bias peaks in the tunneling conductance of spin-orbit-coupled superconducting wires with and without Majorana end-states. Phys. Rev. Lett. **109**(26) (2012). https://doi.org/10.1103/PhysRevLett.109.267002
86. S.M. Albrecht et al., Exponential protection of zero modes in Majorana islands. Nature **531**(7593), 206–209 (2016). https://doi.org/10.1038/nature17162
87. Y.V. Nazarov, Y.M. Blanter, *Quantum Transport: Introduction to Nanoscience* (Cambridge University Press, Cambridge, 2009)
88. C.K. Chiu, J.D. Sau, S. Das Sarma, Conductance of a superconducting coulomb-blockaded Majorana nanowire. Phys. Rev. B **96**(5), 1–23 (2017). https://doi.org/10.1103/PhysRevB.96.054504
89. H.-J. Kwon, V.M. Yakovenko, K. Sengupta, Fractional ac Josephson effect in unconventional superconductors. Low Temp. Phys. **30**(7), 613–619 (2004). https://doi.org/10.1063/1.1789931
90. L.P. Rokhinson, X. Liu, J.K. Furdyna, The fractional a.c. Josephson effect in a semiconductor–superconductor nanowire as a signature of Majorana particles. Nat. Phys. **8**(11), 795–799 (2012). https://doi.org/10.1038/nphys2429
91. J. Wiedenmann et al., 4π-periodic Josephson supercurrent in HgTe-based topological Josephson junctions. Nat. Commun. **7**(1), 1–7 (2016). https://doi.org/10.1038/ncomms10303
92. J.D. Jackson, *Classical Electrodynamics* (Wiley, New York, 1999)
93. A. Fornieri et al., Evidence of topological superconductivity in planar Josephson junctions. Nature **569**(7754), 89–92 (2019). https://doi.org/10.1038/s41586-019-1068-8

94. A. Altland, Y. Gefen, G. Montambaux, What is the Thouless energy for ballistic systems? Phys. Rev. Lett. **76**(7), 1130–1133 (1995). https://doi.org/10.1103/PhysRevLett.76.1130
95. H. Ren et al., Topological superconductivity in a phase-controlled Josephson junction. Nature **569**(7754), 93–98 (2019). https://doi.org/10.1038/s41586-019-1148-9. Nature Publishing Group
96. D. Laroche et al., Observation of the 4π-periodic Josephson effect in indium arsenide nanowires. Nat. Commun. **10**(1), 245 (2019). https://doi.org/10.1038/s41467-018-08161-2
97. K. Le Calvez et al., Joule overheating poisons the fractional ac Josephson effect in topological Josephson junctions. Commun. Phys. **2**(1) (2019). https://doi.org/10.1038/s42005-018-0100-x
98. C.B. Whan, C.J. Lobb, M.G. Forrester, Effect of inductance in externally shunted Josephson tunnel junctions. J. Appl. Phys. **77**(1), 382–389 (1995). https://doi.org/10.1063/1.359334
99. R.F. Miracky, J. Clarke, R.H. Koch, Chaotic noise observed in a resistively shunted self-resonant Josephson tunnel junction. Phys. Rev. Lett. **50**(11), 856–859 (1983). https://doi.org/10.1103/PhysRevLett.50.856
100. A.G. Sun, D.A. Gajewski, M.B. Maple, R.C. Dynes, Observation of Josephson pair tunneling between a high-Tc cuprate (YBa2Cu3O7 $-$ δ) and a conventional superconductor (Pb). Phys. Rev. Lett. **72**(14), 2267–2270 (1994). https://doi.org/10.1103/PhysRevLett.72.2267
101. C.K. Chiu, S. Das Sarma, Fractional Josephson effect with and without Majorana zero modes. Phys. Rev. B **99**(3), 1–13 (2019). https://doi.org/10.1103/PhysRevB.99.035312
102. A. Vuik, B. Nijholt, A.R. Akhmerov, M. Wimmer, *Reproducing Topological Properties with Quasi-Majorana States* (2018)
103. P.W. Anderson, *The Theory of Superconductivity in the High-Tc Cuprates* (Princeton University Press, Princeton, 1997)
104. S. Nadj-Perge et al., Topological matter. Observation of Majorana fermions in ferromagnetic atomic chains on a superconductor. Science **346**(6209), 602–607 (2014). https://doi.org/10.1126/science.1259327
105. D.-J. Choi, N. Lorente, J. Wiebe, K. Von Bergmann, A.F. Otte, A.J. Heinrich, Colloquium: Atomic spin chains on surfaces. Rev. Mod. Phys. **91** (2019). https://doi.org/10.1103/RevModPhys.91.041001
106. J.-X. Zhu, *Bogoliubov-de Gennes Method and its Applications* (Springer, Berlin/Heidelberg, 2016)
107. T.M. Klapwijk, G.E. Blonder, M. Tinkham, Explanation of subharmonic energy gap structure in superconducting contacts. Phys. B+C **109–110**, 1657–1664 (1982)
108. M. Ruby, F. Pientka, Y. Peng, F. Von Oppen, B.W. Heinrich, K.J. Franke, End states and subgap structure in proximity-coupled chains of magnetic adatoms. Phys. Rev. Lett. **115**(19), 197204 (2015). https://doi.org/10.1103/PhysRevLett.115.197204
109. R. Pawlak et al., Probing atomic structure and Majorana wavefunctions in mono-atomic Fe chains on superconducting Pb surface. npj Quant. Infor. **2** (2016). https://doi.org/10.1038/npjqi.2016.35
110. J. Li, H. Chen, I.K. Drozdov, A. Yazdani, B.A. Bernevig, A.H. Macdonald, Topological superconductivity induced by ferromagnetic metal chains. Phys. Rev. B **90**, 235433 (2014). https://doi.org/10.1103/PhysRevB.90.235433
111. Y. Peng, F. Pientka, L.I. Glazman, F. Von Oppen, Strong localization of Majorana end states in chains of magnetic adatoms. Phys. Rev. Lett. **114**(10) (2015). https://doi.org/10.1103/PhysRevLett.114.106801
112. M. Ruby, B.W. Heinrich, Y. Peng, F. Von Oppen, K.J. Franke, Exploring a proximity-coupled co chain on Pb(110) as a possible Majorana platform. Nano Lett. **17**, 44 (2017). https://doi.org/10.1021/acs.nanolett.7b01728
113. B.E. Feldman et al., *High-Resolution Studies of the Majorana Atomic Chain Platform* (2017). https://doi.org/10.1038/NPHYS3947
114. L. Yu, Bound state in superconductors with paramagnetic impurities. Acta Phys. Sin. **21**, 75–91 (1965). https://doi.org/10.7498/aps.21.75

115. A.I. Rusinov, *On the Theory of Gapless Superconductivity in Alloys Containing Paramagnetic Impurities* (1969)
116. J. Li, S. Jeon, Y. Xie, A. Yazdani, B.A. Bernevig, Majorana spin in magnetic atomic chain systems. Phys. Rev. B **97**(12) (2018). https://doi.org/10.1103/PhysRevB.97.125119
117. S. Jeon, Y. Xie, J. Li, Z. Wang, B.A. Bernevig, A. Yazdani, Distinguishing a Majorana zero mode using spin-resolved measurements. Science (80-.). **358**(6364), 772–776 (2017). https://doi.org/10.1126/science.aan3670
118. Y. Zhou, P. Bandaru, D. Sievenpiper, Topologically protected pseudospins in 2D spring-mass system, arXiv:1809.01831 (2018). Accessed on 07 Dec 2018. [Online]. Available: http://arxiv.org/abs/1809.01831
119. W.A. Benalcazar, B.A. Bernevig, T.L. Hughes, Quantized electric multipole insulators. Science **357**(6346), 61–66 (2017). https://doi.org/10.1126/science.aah6442
120. W.A. Benalcazar, B.A. Bernevig, T.L. Hughes, Electric multipole moments, topological multipole moment pumping, and chiral hinge states in crystalline insulators. Phys. Rev. B **96**(24), 245115 (2017). https://doi.org/10.1103/PhysRevB.96.245115
121. Z. Song, Z. Fang, C. Fang, (d − 2) -dimensional edge states of rotation symmetry protected topological states. Phys. Rev. Lett. **119**(24), 246402 (2017). https://doi.org/10.1103/PhysRevLett.119.246402
122. M. Ezawa, Higher-order topological insulators and semimetals on the breathing Kagome and Pyrochlore lattices. Phys. Rev. Lett. **120**(2), 026801 (2018). https://doi.org/10.1103/PhysRevLett.120.026801
123. H. Xue, Y. Yang, F. Gao, Y. Chong, B. Zhang, Acoustic higher-order topological insulator on a kagome lattice. Nat. Mater. **18**(2), 108–112 (2019). https://doi.org/10.1038/s41563-018-0251-x
124. X. Ni, M. Weiner, A. Alu, A.B. Khanikaev, Observation of higher order topological acosutic states protected by generalized chiral symmetry. Nat. Mater. **18**, 113–120 (2019)
125. B. Jäck, Y. Xie, J. Li, S. Jeon, B.A. Bernevig, A. Yazdani, Observation of a Majorana zero mode in a topologically protected edge channel. Science **364**(6447), 1255–1259 (2019). https://doi.org/10.1126/science.aax1444
126. F. Schindler et al., Higher-order topology in bismuth. Nat. Phys. **14**(9), 918–924 (2018). https://doi.org/10.1038/s41567-018-0224-7
127. I.K. Drozdov et al., One-dimensional topological edge states of bismuth bilayers. Nat. Phys. **10**(9), 664–669 (2014). https://doi.org/10.1038/nphys3048
128. X. Sun et al., in *18th ICT Symposium Proceedings*, Experimental study of the effect of the quantum well structures on the thermoelectric figure of merit in Si/Si1-xGex system (1999), pp. 652–655
129. P. Pichanusakorn, P.R. Bandaru, Nanostructured thermoelectrics. Mater. Sci. Eng. R **67**, 19–63 (2010)
130. Z.F. Ezawa, *Quantum Hall Effects: Recent Theoretical and Experimental Developments* (World Scientific Publishing Company, 2013)
131. N. Read, D. Green, Paired states of fermions in two dimensions with breaking of parity and time-reversal symmetries and the fractional quantum hall effect. Phys. Rev. B – Condens. Matter Mater. Phys. **61**(15), 10267–10297 (2000). https://doi.org/10.1103/PhysRevB.61.10267
132. C.-X. Liu, S.-C. Zhang, X.-L. Qi, The quantum anomalous hall effect: Theory and experiment. Annu. Rev. Condens. Matter Phys. **7**(1), 301–321 (2016). https://doi.org/10.1146/annurev-conmatphys-031115-011417
133. C.Z. Chang et al., Experimental observation of the quantum anomalous hall effect in a magnetic topological insulator. Science (80-.) **340**(6129), 167–170 (2013). https://doi.org/10.1126/science.1234414
134. Q.L. He et al., Chiral Majorana fermion modes in a quantum anomalous hall insulator–superconductor structure. Science (80-.) **357**(6348), 294–299 (2017). https://doi.org/10.1126/science.aag2792

135. X. Kou, Y. Fan, M. Lang, P. Upadhyaya, K.L. Wang, *Magnetic Topological Insulators and Quantum Anomalous Hall Effect* (2014). https://doi.org/10.1016/j.ssc.2014.10.022
136. C.-Z. Chang, M. Li, *Quantum Anomalous Hall Effect in Time-Reversal-Symmetry Breaking Topological Insulators* (2016). https://doi.org/10.1088/0953-8984/28/12/123002
137. Y. Tokura, K. Yasuda, A. Tsukazaki, Magnetic topological insulators. Nat. Rev. Phys. **1**(2), 126–143 (2019). https://doi.org/10.1038/s42254-018-0011-5
138. C.W.J. Beenakker, P. Baireuther, Y. Herasymenko, I. Adagideli, L. Wang, A.R. Akhmerov, Deterministic creation and braiding of chiral edge vortices. Phys. Rev. Lett. **122** (2019). https://doi.org/10.1103/PhysRevLett.122.146803
139. B. Lian, X.Q. Sun, A. Vaezi, X.L. Qib, S.C. Zhang, Topological quantum computation based on chiral Majorana fermions. Proc. Natl. Acad. Sci. U. S. A. **115**(43), 10938–10942 (2018). https://doi.org/10.1073/pnas.1810003115
140. R. Yu, W. Zhang, H.-J. Zhang, S.-C. Zhang, X. Dai, Z. Fang, Quantized anomalous hall effect in magnetic topological insulators. Science (80-.) **329**, 61 (2010)
141. Q.L. He et al., Chiral Majorana fermion modes in a quantum anomalous hall insulator-superconductor structure. Science **357**(6348), 294–299 (2017). https://doi.org/10.1126/science.aag2792
142. X.L. Qi, T.L. Hughes, S.C. Zhang, Topological field theory of time-reversal invariant insulators. Phys. Rev. B – Condens. Matter Mater. Phys. **78**(19) (2008). https://doi.org/10.1103/PhysRevB.78.195424
143. W. Ji, X.-G. Wen, 1/2 (e^2 /h) conductance plateau without 1D Chiral Majorana fermions. Phys. Rev. Lett. **120**, 107002 (2018). https://doi.org/10.1103/PhysRevLett.120.107002
144. M. Kayyalha et al., Absence of evidence for chiral Majorana modes in quantum anomalous hall-superconductor devices. Science (80-.). **367**(6473), 64–67 (2020). https://doi.org/10.1126/science.aax6361
145. Y. Huang, F. Setiawan, J.D. Sau, Disorder-induced half-integer quantized conductance plateau in quantum anomalous hall insulator-superconductor structures. Phys. Rev. B **97**, 100501 (2018). https://doi.org/10.1103/PhysRevB.97.100501
146. P. Hosur, P. Ghaemi, R.S.K. Mong, A. Vishwanath, Majorana modes at the ends of superconductor vortices in doped topological insulators. Phys. Rev. Lett. **107**(9), 097001 (2011). https://doi.org/10.1103/PhysRevLett.107.097001
147. T.A. Fulton, R.C. Dynes, P.W. Andesron, The flux shuttle—A Josephson junction shift register employing single flux quanta. Proc. IEEE **61**, 28 (1973)
148. K.K. Likharev, V.K. Semenov, RSFQ logic/memory family: A new Josephson-junction technology for sub-terahertz-clock-frequency digital systems. IEEE Trans. Appl. Supercond. **1**, 3 (1991)
149. H.F. Hess, R.B. Robinson, R.C. Dynes, J.M. Valles, J.V. Waszczak, Scanning-tunneling microscope observation of the Abrikosov flux lattice and the density of states near and inside a fluxoid. Phys. Rev. Lett. **62**, 214–216 (1989)
150. C. Caroli, P.-G. de Gennes, J. Matricon, Bound fermion states on a vortex line in a type II superconductor. Phys. Lett. **9**, 307–309 (1964)
151. J.-P. Xu et al., Artificial topological superconductor by the proximity effect. Phys. Rev. Lett. **112**(21), 217001 (2014). https://doi.org/10.1103/PhysRevLett.112.217001
152. S. Zhu et al., *Observation of Majorana Conductance Plateau by Scanning Tunneling Spectroscopy* (2019, April). Accessed 25 Jul 2019. [Online]. Available: http://arxiv.org/abs/1904.06124.
153. J.-P. Xu et al., Experimental detection of a Majorana mode in the core of a magnetic vortex inside a topological insulator-superconductor Bi2Te3 /NbSe2 Heterostructure. Phys. Rev. Lett. **114**(1), 017001 (2015). https://doi.org/10.1103/PhysRevLett.114.017001
154. H.-H. Sun et al., Majorana zero mode detected with spin selective Andreev reflection in the vortex of a topological superconductor. Phys. Rev. Lett. **116**(25), 257003 (2016). https://doi.org/10.1103/PhysRevLett.116.257003

155. N. Djavid, G. Yin, Y. Barlas, R.K. Lake, Gate controlled Majorana zero modes of a two-dimensional topological superconductor. Appl. Phys. Lett. **113**(1), 012601 (2018). https://doi.org/10.1063/1.5027440
156. P.-G. de Gennes, *Superconductivity of Metals and Alloys* (Westview Press, 1966)
157. M.Z. Hasan, C.L. Kane, Colloquium: Topological insulators. Rev. Mod. Phys. **82**(4), 3045–3067 (2010). https://doi.org/10.1103/RevModPhys.82.3045
158. D. Wang et al., Evidence for Majorana bound states in an iron-based superconductor. Science **362**(6412), 333–335 (2018). https://doi.org/10.1126/science.aao1797
159. F.C. Hsu et al., Superconductivity in the PbO-type structure α-FeSe. Proc. Natl. Acad. Sci. U. S. A. **105**(38), 14262–14264 (2008). https://doi.org/10.1073/pnas.0807325105
160. Q. Liu et al., Robust and clean Majorana zero mode in the vortex Core of high-temperature superconductor (Li0.84Fe0.16)OHFeSe. Phys. Rev. X **8**(4), 41056 (2018). https://doi.org/10.1103/PhysRevX.8.041056
161. J.-X. Yin et al., Observation of a robust zero-energy bound state in iron-based superconductor Fe(Te,Se). Nat. Phys. **11**(7), 543–546 (2015). https://doi.org/10.1038/nphys3371
162. D. Wang, J. Wiebe, R. Zhong, G. Gu, R. Wiesendanger, Spin-polarized Yu-Shiba-Rusinov states in an iron-based superconductor. Phys. Rev. Lett. **126** (2021). https://doi.org/10.1103/PhysRevLett.126.076802
163. P. Yu et al., Non-Majorana states yield nearly quantized conductance in proximatized nanowires. Nat. Phys. **17**(4), 482–488 (2021). https://doi.org/10.1038/S41567-020-01107-W
164. E.J.H. Lee, X. Jiang, M. Houzet, R. Aguado, C.M. Lieber, S. De Franceschi, Spin-resolved Andreev levels and parity crossings in hybrid superconductor-semiconductor nanostructures. Nat. Nanotechnol. **9**(1), 79–84 (2014). https://doi.org/10.1038/NNANO.2013.267
165. J. C. Estrada Saldaña et al., *Bias Asymmetric Subgap States Mimicking Majorana Signatures*
166. P.W. Brouwer, M. Duckheim, A. Romito, F. Von Oppen, Probability distribution of Majorana end-state energies in disordered wires. Phys. Rev. Lett. **107**(19), 196804 (2011). https://doi.org/10.1103/PhysRevLett.107.196804
167. S.M. Ross, *Introduction to Probability Models*, 14th edn. (Academic Press, New York, 2014)
168. A. Haim, A. Stern, Benefits of weak disorder in one-dimensional topological superconductors. Phys. Rev. Lett. **122**(12), 126801 (2019). https://doi.org/10.1103/PhysRevLett.122.126801
169. K.A. Kouznetsov et al., *C-Axis Josephson Tunneling Between YBCO and Pb: Direct Evidence for Mixed Order Parameter Symmetry in a High-T_c Superconductor* (1997, May). https://doi.org/10.1103/physrevlett.79.3050
170. http://public.itrs.net/, *International Technology Roadmap for Semiconductors* (2017)
171. M. Valentini et al., Nontopological zero-bias peaks in full-shell nanowires induced by flux-tunable Andreev states. Science (80-.) **373**(6550), 82–88 (2021). https://doi.org/10.1126/SCIENCE.ABF1513
172. H. Hafiz et al., A high-throughput data analysis and materials discovery tool for strongly correlated materials. npj Comput. Mater. **4**(1), 63 (2018). https://doi.org/10.1038/s41524-018-0120-9
173. S. Sasaki et al., Topological superconductivity in CuxBi2Se3. Phys. Rev. Lett. **107**, 217001 (2011)
174. Q. Liu et al., Robust and clean Majorana zero mode in the vortex core of high-temperature superconductor (Li0.84Fe0.16)OHFeSe. Phys. Rev. X **8**, 041056 (2018)
175. H. Kimura, R.P. Barber Jr., S. Ono, Y. Ando, R.C. Dynes, Scanning Josephson tunneling microscopy of single crystal Bi2Sr2CaCu2O8+d with a conventional superconducting tip. Phys. Rev. Lett. **101**, 037002 (2008)
176. H. Choi et al., Experimental and theoretical study of topology and electronic correlations in PuB 4. Phys. Rev. B **97**(20), 201114 (2018). https://doi.org/10.1103/PhysRevB.97.201114
177. E.Y. Cho, Y.W. Zhou, J.Y. Cho, S.A. Cybart, Superconducting nano Josephson junctions patterned with a focused helium ion beam. Appl. Phys. Lett. **113**, 022604 (2018)
178. S. Das Sarma, M. Freedman, C. Nayak, Majorana zero modes and topological quantum computation. npj Quantum Inf. **1**, 15001 (2015)

179. A.K. Niessen, J. van Suchtelen, F.A. Staas, W.F. Druyvesteyn, Guided motion of vortices and anisotropic resistivity in Type-II superconductors. Philips Res. Reports **20**, 226–234 (1965)
180. P. Nozieres, W.F. Vinen, The motion of flux lines in Type II superconductors. Philos. Mag. **14**, 667 (1966)
181. S.A. Cybart et al., Temporal Stability of Y–Ba–Cu–O Nano Josephson Junctions from Ion Irradiation. IEEE Trans. Appl. Supercond. **23**, 1100103 (2013)
182. S.A. Cybart et al., Comparison of Y–Ba–Cu–O films irradiated with helium and neon ions for the fabrication of Josephson devices. IEEE Trans. Appl. Supercond. **24**, 1100105 (2014)
183. C.-X. Liu, D. E. Liu, F.-C. Zhang, and C.-K. Chiu, Protocol for reading out Majorana vortex qubit and testing non-Abelian statistics (2019). Accessed 24 Jul 2019. [Online]. Available: https://arxiv.org/abs/1901.06083
184. P. Bonderson, M. Freedman, C. Nayak, Measurement-only topological quantum computation. Phys. Rev. Lett. **101**(1), 010501 (2008). https://doi.org/10.1103/PhysRevLett.101.010501
185. S. Plugge, A. Rasmussen, R. Egger, K. Flensberg, Majorana box qubits. New J. Phys. **19**(1), 012001 (2017). https://doi.org/10.1088/1367-2630/aa54e1
186. M.J.A. Schuetz, E.M. Kessler, G. Giedke, L.M.K. Vandersypen, M.D. Lukin, J.I. Cirac, Universal quantum transducers based on surface acoustic waves. Phys. Rev. X **5**, 031031 (2015)
187. M. Weiss, H.J. Krenner, Interfacing quantum emitters with propagating surface acoustic waves. J. Phys. D. Appl. Phys. **51**, 373001 (2018)
188. D. Aasen et al., Milestones toward Majorana-based quantum computing. Phys. Rev. X **6**(3), 031016 (2016). https://doi.org/10.1103/PhysRevX.6.031016
189. R.S. Keizer, S.T.B. Goennenwein, T.M. Klapwijk, G. Miao, G. Xiao, A. Gupta, A spin triplet supercurrent through the half-metallic ferromagnet CrO 2. Nature **439**(7078), 825–827 (2006). https://doi.org/10.1038/nature04499
190. F.S. Bergeret, A.F. Volkov, K.B. Efetov, Odd triplet superconductivity and related phenomena in superconductor- ferromagnet structures. Rev. Mod. Phys. **77**(4), 1321–1373 (2005). https://doi.org/10.1103/RevModPhys.77.1321
191. S. Frolov, Quantum computing's reproducibility crisis: Majorana fermions. Nature **592**(7854), 350–352 (2021). https://doi.org/10.1038/d41586-021-00954-8
192. Q. Zhou et al., Chiral Majorana fermion modes in a quantum anomalous hall insulator–superconductor structure. Science (80-.) **357**(6348), 294–299 (2017). https://doi.org/10.1126/science.aag2792
193. S.M. Frolov, M.J. Manfra, J.D. Sau, Topological superconductivity in hybrid devices. Nat. Phys. **16**(7), 718–724 (2020). https://doi.org/10.1038/S41567-020-0925-6

Index

© The Author(s), under exclusive license to Springer Nature Switzerland AG 2022

P. Bandaru, S. Natani, *Topological States for New Modes of Information Storage and Transfer*, https://doi.org/10.1007/978-3-030-93340-1

The manufacturer's authorised representative in the EU is Springer Nature Customer Service Centre GmbH, Europaplatz 3, 69115 Heidelberg, Germany. If you have any concerns regarding our products, please contact ProductSafety@springernature.com

Printed and bound by CPI Group (UK) Ltd, Croydon, CR0 4YY
30/09/2026
02225710-0005